Sincerely,
G.R. Partridge
1990

GIFTED HANDS

GIFTED HANDS

Ben Carson, M.D., with Cecil Murphey

REVIEW AND HERALD® PUBLISHING ASSOCIATION
WASHINGTON, DC 20039-0555
HAGERSTOWN, MD 21740

The authors assume full responsibility for the accuracy of all facts and
quotations as cited in this book.

This book was
Edited by Penny Estes Wheeler
Designed by Bill Kirstein
Cover photo by Christine Armstrong
Type set: 11/12 Garamond Book

Bible texts credited to RSV are from the Revised Standard Version of the
Bible, copyrighted 1946, 1952 © 1971, 1973.

PRINTED IN U.S.A.

94 93 92 91 90   10 9 8 7 6 5 4 3

Library of Congress Cataloging in Publication Data

Carson, Benjamin S.
    Gifted hands / Ben Carson, with Cecil Murphey.
        p.   cm.
        1. Carson, Benjamin S. 2. Neurosurgeons—United States—Biography.
I. Murphey, Cecil B.   II. Title.
[RD592.9.C37A3 1990b]
617.4′8092—dc20
[B]   90-8326
                                                                          CIP

ISBN 0-8280-0561-3

Dedication

*This book
is dedicated to my mother,
SONYA CARSON,
who basically sacrificed her life
to make certain that my brother and I
got a head start.*

Contents

CHAPTER 1
"Goodbye, Daddy"/11

CHAPTER 2
Carrying the Load/17

CHAPTER 3
Eight Years Old/23

CHAPTER 4
Two Positives/32

CHAPTER 5
A Boy's Big Problem/45

CHAPTER 6
A Terrible Temper/54

CHAPTER 7
ROTC Triumph/61

CHAPTER 8
College Choices/71

CHAPTER 9
Changing the Rules/80

CHAPTER 10
A Serious Step/91

CHAPTER 11
Another Step Forward/104

CHAPTER 12
Coming Into My Own/123

CHAPTER 13
A Special Year/135

CHAPTER 14
A Girl Named Maranda/146

CHAPTER 15
Heartbreak/155

CHAPTER 16
Little Beth/167

CHAPTER 17
Three Special Children/177

CHAPTER 18
Craig and Susan/185

CHAPTER 19
Separating the Twins/201

CHAPTER 20
The Rest of Their Story/213

CHAPTER 21
Family Affairs/219

CHAPTER 22
Think Big/225

Introduction

by Candy Carson

"More blood! Stat!"

The silence of the OR was smashed by the amazingly quiet command. The twins had received 50 units of blood, but their bleeding still hadn't stopped!

"There's no more type-specific blood," the reply came. "We've used it all."

As a result of this announcement, a quiet panic erupted through the room. Every ounce of type AB* negative blood had been drained from the Johns Hopkins Hospital blood bank. Yet the 7-month-old twin patients who had been joined at the back of their heads since birth needed more blood or they would die without ever having a chance to recuperate. This was their only opportunity, their only chance, at normal lives.

Their mother, Theresa Binder, had searched throughout the medical world and found only one team who was willing to even attempt to separate her twin boys *and* preserve both lives. Other surgeons told her it couldn't be done—that one of the boys would have to be sacrificed. *Allow one of her darlings to die?* Theresa couldn't even bear the thought. Although they were joined at the head, even at 7 months of age each had his own personality—one playing while the other slept or ate. No, she absolutely couldn't do it! After months of searching she discovered the Johns Hopkins team.

Many of the 70-member team began offering to donate their own blood, realizing the urgency of the situation.

The 17 hours of laborious, tedious, painstaking operating on such tiny patients had progressed well, all things considered. The babies had been successfully anesthetized after only a few hours, a complex procedure because of their shared blood vessels. The preparation for cardiovascular bypass hadn't taken much longer than expected (the five months of planning and numerous dress rehearsals had paid off). Getting to the site of the twins' juncture wasn't particularly difficult for the young, though seasoned, neurosurgeons either. But, as a result of the cardiovascular bypass procedures, the blood lost its clotting properties. Therefore, every place in the infants' heads that could bleed did bleed!

Fortunately, within a short time the city blood bank was able to locate the exact number of units of blood needed to continue the surgery. Using every skill, trick, and device known in their specialities, the surgeons were able to stop the bleeding within a couple of hours. The operation continued. Finally, the plastic surgeons sewed the last skin flaps to close the wounds, and the 22-hour surgical ordeal was over. The Siamese twins —Patrick and Benjamin—were separate for the first time in their lives!

The exhausted primary neurosurgeon who had devised the plan for the operation was a ghetto kid from the streets of Detroit.

---

* Blood type changed for privacy.

1 "GOODBYE, DADDY"

And your daddy isn't going to live with us anymore."

"Why not?" I asked again, choking back the tears. I just could not accept the strange finality of my mother's words. "I love my dad!"

"He loves you too, Bennie . . . but he has to go away. For good."

"But why? I don't want him to go. I want him to stay here with us."

"He's got to go—"

"Did I do something to make him want to leave us?"

"Oh, no, Bennie. Absolutely not. Your daddy loves you."

I burst into tears. "Then make him come back."

"I can't. I just can't." Her strong arms held me close, trying to comfort me, to help me stop crying. Gradually my sobs died away, and I calmed down. But as soon as she loosened her hug and let me go, my questions started again.

"Your Daddy did—" Mother paused, and, young as I was, I knew she was trying to find the right words to

11

make me understand what I didn't want to grasp. "Bennie, your daddy did some bad things. Real bad things."

I swiped my hand across my eyes. "You can forgive him then. Don't let him go."

"It's more than just forgiving him, Bennie—"

"But I want him to stay here with Curtis and me and you."

Once again Mother tried to make me understand why Daddy was leaving, but her explanation didn't make a lot of sense to me at 8 years of age. Looking back, I don't know how much of the reason for my father's leaving sank into my understanding. Even what I grasped, I wanted to reject. My heart was broken because Mother said that my father was never coming home again. And I loved him.

Dad was affectionate. He was often away, but when he was home he'd hold me on his lap, happy to play with me whenever I wanted him to. He had great patience with me. I particularly liked to play with the veins on the back of his large hands, because they were so big. I'd push them down and watch them pop back up. "Look! They're back again!" I'd laugh, trying everything within the power of my small hands to make his veins stay down. Dad would sit quietly, letting me play as long as I wanted.

Sometimes he'd say, "Guess you're just not strong enough," and I'd push even harder. Of course nothing worked, and I'd soon lose interest and play with something else.

Even though Mother said that Daddy had done some bad things, I couldn't think of my father as "bad," because he'd always been good to my brother, Curtis, and me. Sometimes Dad brought us presents for no special reason. "Thought you'd like this," he'd say offhandedly, a twinkle in his dark eyes.

Many afternoons I'd pester my mother or watch the clock until I knew it was time for my dad to come home from work. Then I'd rush outside to wait for him. I'd watch until I saw him walking down our alley. "Daddy!

Daddy!" I'd yell, running to meet him. He would scoop me into his arms and carry me into the house.

That stopped in 1959 when I was 8 years old and Daddy left home for good. To my young, hurting heart the future stretched out forever. I couldn't imagine a life without Daddy and didn't know if Curtis, my 10-year-old brother, or I would ever see him again.

I don't know how long I continued the crying and questioning the day Daddy left; I only know it was the saddest day of my life. And my questions didn't stop with my tears. For weeks I pounded my mother with every possible argument my mind could conceive, trying to find some way to get her to make Daddy come back home.

"How can we get by without Daddy?"

"Why don't you want him to stay?"

"He'll be good. I know he will. Ask Daddy. He won't do bad things again."

My pleading didn't make any difference. My parents had settled everything before they told Curtis and me.

"Mothers and fathers are supposed to stay together," I persisted. "They're both supposed to be with their little boys."

"Yes, Bennie, but sometimes it just doesn't work out right."

"I still don't see why," I said. I thought of all the things Dad did with us. For instance, on most Sundays, Dad would take Curtis and me for drives in the car. Usually we visited people, and we'd often stop by to see one family in particular. Daddy would talk with the grown-ups, while my brother and I played with the children. Only later did we learn the truth—my father had another "wife" and other children that we knew nothing about.

I don't know how my mother found out about his double life, for she never burdened Curtis and me with the problem. In fact, now that I'm an adult, my one complaint is that she went out of her way to protect us from knowing how bad things were. We were never

13

allowed to share how deeply she hurt. But then, that was Mother's way of protecting us, thinking she was doing the right thing. And many years later I finally understood what she called his "betrayals with women and drugs."

Long before Mother knew about the other family, I sensed things weren't right between my parents. My parents didn't argue; instead, my father just walked away. He had been leaving the house more and more and staying away longer and longer. I never knew why.

Yet when Mother told me "Your daddy isn't coming back," those words broke my heart.

I didn't tell Mother, but every night when I went to bed I prayed, "Dear Lord, help Mother and Dad get back together again." In my heart I just knew God would help them make up so we could be a happy family. I didn't want them to be apart, and I couldn't imagine facing the future without my father.

But Dad never came home again.

As the days and weeks passed, I learned we could get by without him. We were poorer then, and I could tell Mother worried, although she didn't say much to Curtis or me. As I grew wiser, and certainly by the time I was 11, I realized that the three of us were actually happier than we had been with Dad in the house. We had peace. No periods of deathly silence filled the house. I no longer froze with fear or huddled in my room, wondering what was happening when Mother and Daddy didn't talk.

That's when I stopped praying for them to get back together. "It's better for them to stay split up," I said to Curtis. "Isn't it?"

"Yeah, guess so," he answered. And, like Mother, he didn't say much to me about his own feelings. But I think I knew that he too reluctantly realized that our situation was better without our father.

Trying to remember how I felt in those days after Dad left, I'm not aware of going through stages of anger and resentment. My mother says that the experience pushed Curtis and me into a lot of pain. I don't doubt that his leaving meant a terrible adjustment for both of us boys. Yet I still have no recollection beyond his initial leaving.

Maybe that's how I learned to handle my deep hurt—by forgetting.

───

"We just don't have the money, Bennie."

In the months after Dad left, Curtis and I must have heard that statement a hundred times, and, of course, it was true. When we asked for toys or candy, as we'd done before, I soon learned to tell from the expression on Mother's face how deeply it hurt her to deny us. After a while I stopped asking for what I knew we couldn't have anyway.

In a few instances resentment flashed across my mother's face. Then she'd get very calm and explain to us boys that Dad loved us but wouldn't give her any money to support us. I vaguely recall a few times when Mother went to court, trying to get child support from him. Afterward, Dad would send money for a month or two—never the full amount—and he always had a legitimate excuse. "I can't give you all of it this time," he'd say, "but I'll catch up. I promise."

Dad never caught up. After a while Mother gave up trying to get any financial help from him.

I was aware that he wouldn't give her money, which made life harder on us. And in my childish love for a dad who had been kind and affectionate, I didn't hold it against him. But at the same time I couldn't understand how he could love us and not want to give us money for food.

One reason I didn't hold any grudges or harsh feelings toward Dad must have been that my mother seldom blamed him—at least not to us or in our hearing. I can hardly think of a time when she spoke against him.

More important than that fact, though, Mother managed to bring a sense of security to our three-member family. While I still missed Dad for a long time, I felt a sense of contentment being with just my mother and my brother because we really did have a happy family.

My mother, a young woman with hardly any education, came from a large family and had many things against her. Yet she pulled off a miracle in her own life, and helped in ours. I can still hear Mother's voice, no matter how bad things were, saying, "Bennie, we're going to be fine." Those weren't empty words either, for she believed them. And because she believed them, Curtis and I believed them too, and they provided a comforting assurance for me.

Part of Mother's strength came from a deep-seated faith in God and perhaps just as much from her innate ability to inspire Curtis and me to know she meant every word she said. We knew we weren't rich; yet no matter how bad things got for us, we didn't worry about what we'd have to eat or where we'd live.

Our growing up without a father put a heavy burden on my mother. She didn't complain—at least not to us—and she didn't feel sorry for herself. She tried to carry the whole load, and somehow I understood what she was doing. No matter how many hours she had to be away from us at work, I knew she was doing it for us. That dedication and sacrifice made a profound impression on my life.

Abraham Lincoln once said, "All that I am or ever hope to be, I owe to my mother." I'm not sure I want to say it quite like that, but my mother, Sonya Carson, was the earliest, strongest, and most impacting force in my life.

It would be impossible to tell about my accomplishments without starting with my mother's influence. For me to tell my story means beginning with hers.

2 CARRYING THE LOAD

T hey're not going to treat my boy that way," Mother said as she stared at the paper Curtis had given her. "No, sir, they're not going to do that to you." Curtis had had to read some of the words to her, but she understood exactly what the school counselor had done.

"What you going to do, Mother?" I asked in surprise. It had never occurred to me that anyone could change anything when school authorities made decisions.

"I'm going right over there in the morning and get this straightened out," she said. From the tone of her voice I knew she'd do it.

Curtis, two years my senior, was in junior high school when the school counselor decided to place him into the vocational-type curriculum. His once-low grades had been climbing nicely for more than a year, but he was enrolled in a predominantly White school, and Mother had no doubt that the counselor was operating from the stereotypical thinking that Blacks were incapable of college work.

Of course, I wasn't at their meeting, but I still vividly

remember what Mother told us that evening. "I said to that counselor woman, 'My son Curtis is going to college. I don't want him in any vocational courses.'" Then she put her hand on my brother's head. "Curtis, you are now in the college prep courses."

That story illustrates my mother's character. She was not a person who would allow the system to dictate her life. Mother had a clear understanding of how things would be for us boys.

My mother is an attractive woman, five feet three and slim, although when we were kids I'd say she was on the plump side of medium. Today she suffers from arthritis and heart problems, but I don't think she has slowed down much.

Sonya Carson has the classic Type A personality —hardworking, goal-oriented, driven to demanding the best of herself in any situation, refusing to settle for less. She's highly intelligent, a woman who quickly grasps the overall significance rather than searching for details. She has a natural ability—an intuitive sense—that enables her to perceive what should be done. That's probably her most outstanding characteristic.

Because of that determined, perhaps compulsive, personality that demanded so much from herself, she infused some of that spirit into me. I don't want to portray my mother as perfect because she was human too. At times her refusing to allow me to settle for less than the best came across as nagging, demanding, even heartless to me. When she believed in something she held on and wouldn't quit. I didn't always like hearing her say, "You weren't born to be a failure, Bennie. You can do it!" Or one of her favorites: "You just ask the Lord, and He'll help you."

Being kids, we didn't always welcome her lessons and advice. Resentment and obstinance crept in, but my mother refused to give up.

Over a period of years, with Mother's constant encouragement, both Curtis and I started believing that we really could do anything we chose to do. Maybe she brainwashed us into believing that we were going to be extremely good and highly successful at whatever we attempted. Even today I can clearly hear her voice in the back of my head saying, "Bennie, you can do it. Don't you stop believing that for one second."

Mother had only a third-grade education when she married, yet she provided the driving force in our home. She pushed my laid-back father to do a lot of things. Largely because of her sense of frugality, they saved a fair amount of money and eventually bought our first house. I suspect that, had things gone Mother's way, ultimately they would have been financially well-off. And I'm sure she had no premonition of the poverty and hardship she'd have to face in the years ahead.

By contrast, my father was six feet two, slender, and he often said, "You got to look sharp all the time, Bennie. Dress the way you want to be." He emphasized clothes and possessions, and he enjoyed being around people.

"Be nice to people. People are important, and if you're nice to them, they'll like you." Recalling those words, I believe he put great importance on being liked by everybody. If anyone asked me to describe my dad, I'd have to say, "He's just a nice guy." And, despite all the problems that erupted later, I feel that way today.

My father was the kind of person who would have wanted us to wear the fancy clothes and to do the macho kind of things like girl hunting—the lifestyle that would have been detrimental to establishing ourselves academically. In many ways, I'm now grateful my mother took us out of that environment.

Intellectually, Dad didn't easily grasp complex problems because he tended to get bogged down in details, unable to see the overall picture. That was probably the biggest difference between my parents.

Both parents came from big families: my mother had 23 siblings, and my father grew up with 13 brothers and sisters. They married when my father was 28 and my mother was 13. Many years later she confided that she was looking for a way to get out of a desperate home situation.

Shortly after their marriage, they moved from Chattanooga, Tennessee, to Detroit, which was the trend for laborers in the late 1940s and early 1950s. People from the rural South migrated toward what they considered lucrative factory jobs in the North. My father got a job at the Cadillac plant. So far as I know, it was the first and only employment he ever held. He worked for Cadillac until he retired in the late 1970s.

My father also served as a minister in a small Baptist church. I've never been able to understand whether he was an ordained minister or not. Only one time did Daddy take me to hear him preach—or at least I remember only one occasion. Daddy wasn't one of those fiery types like some television evangelists. He spoke rather calmly, raised his voice a few times, but he preached in a relatively low key, and the audience didn't get stirred up. He didn't have a real flow of words, but he did the best he could. I can still see him on that special Sunday as he stood in front of us, tall and handsome, the sun glinting off a large metal cross that dangled across his chest.

———

"I'm going away for a few days," Mother said several months after Daddy left us. "Going to see some relatives."

"We going too?" I asked with interest.

"No, I have to go alone." Her voice was unusually quiet. "Besides, you boys can't miss school."

Before I could object, she told me that we could stay with neighbors. "I've already arranged it for you. You can sleep over there and eat with them until I come back."

Maybe I should have wondered why she left, but I didn't. I was so excited to stay in somebody else's house because that meant extra privileges, better food, and a lot of fun playing with the neighbor kids.

That's the way it happened the first time and several times after that. Mother explained that she was going away for a few days, and we would be taken care of by our neighbors. Because she carefully arranged for us to stay with friends, it was exciting rather than frightful. Secure in her love, it never occurred to me that she wouldn't be back.

It may seem strange, but it is a testimony to the security we felt in our home—I was an adult before I discovered where Mother went when she "visited relatives." When the load became too heavy, she checked herself into a mental institution. The separation and divorce plunged her into a terrible period of confusion and depression, and I think her inner strength helped her realize she needed professional help and gave her the courage to get it. Usually she was gone for several weeks at a time.

We boys never had the slightest suspicion about her psychiatric treatment. She wanted it that way.

With time, Mother rebounded from her mental pressures, but friends and neighbors found it hard to accept her as healthy. We kids never knew it, for Mother never let on how it hurt her, but her treatment in a mental hospital provided neighbors with a hot topic of gossip, perhaps even more because she had gone through a divorce. Both problems created serious stigmas at the time. Mother not only had to cope with providing a home and making a living to support us, but most of her friends disappeared when she needed them most.

Because Mother never talked to anyone about the details of her divorce, people assumed the worst and circulated wild stories about her.

"I just decided that I had to go about my own business," Mother once told me, "and ignore what people said." She did, but it couldn't have been easy. It hurts to think of how many lonely, tearful times she suffered alone.

Finally, with no financial resources to fall back on, Mother knew she couldn't keep up the expenses of living in our house, modest as it was. The house was hers, as part of the divorce settlement. So after several months of trying to make it on her own, Mother rented out the house, packed us up, and we moved away. This was one of the times when Dad reappeared, for he came back to drive us to Boston. Mother's older sister, Jean Avery, and her husband, William, agreed to take us in.

We moved into the Boston tenements with the Averys. Their children were grown, and they had a lot of love to share with two little boys. In time, they became like another set of parents to Curtis and me, and that was wonderful for we needed a lot of affection and sympathy then.

For a year or so after we moved to Boston, Mother still underwent psychiatric treatment. Her trips away lasted three or four weeks each time. We missed her, but we received such special attention from Uncle William and Aunt Jean when she was gone that we liked the occasional arrangement.

The Averys assured Curtis and me, "Your mama is doing just fine." After getting a letter or a telephone call they'd tell us, "She'll be back in a few more days." They handled the situation so well that we never had any idea how tough things were for our mother. And that's just how the strong-willed Sonya Carson wanted it to be.

3 EIGHT YEARS OLD

Rats!" I yelled. "Hey, Curt, lookey there! I saw rats!" I pointed in horror to a large weeded area behind our tenement building. "And they're bigger than cats!"

"Not quite that big," Curtis countered, trying to sound more mature. "But they sure are mean-looking."

Nothing in Detroit had prepared us for life in a Boston tenement. Armies of roaches streaked across the room, impossible to get rid of no matter what Mother did. More frightening to me were the hordes of rats, even though they never got close. Mostly they lived outside in the weeds or piles of debris. But occasionally they scurried into the basement of our building, especially during the cold weather.

"I'm not going down there by myself," I said adamantly more than once. I was scared to go down into the basement alone. And I wouldn't budge unless Curtis or Uncle William went with me.

Sometimes snakes came out of the weeds to slither down the sidewalks. Once a big snake crawled into our basement, and someone killed it. For days afterward all us kids talked about snakes.

"You know, a snake got into one of those buildings behind us last year and killed four children in their sleep," one of my classmates said.

"They gobble you up," insisted another.

"No, they don't," the first one said and laughed. "They kind of sting you and then you die." Then he told another story about somebody being killed by a snake.

The stories weren't true, of course, but hearing them often enough kept them in my mind, making me cautious, fearful, and always on the lookout for snakes.

A lot of winos and drunks flopped around the area, and we became so used to seeing broken glass, trashed lots, dilapidated buildings, and squad cars racing up the street that we soon adjusted to our change of lifestyle. Within weeks this setting seemed perfectly normal and reasonable.

No one ever said, "This isn't the way normal people live." Again, I think it was the sense of family unity, strengthened by the Averys, that kept me from being too concerned about the quality of our life in Boston.

Of course, Mother worked. Constantly. She seldom had much free time, but she showered that time on Curtis and me, which made up for the hours she was away. Mother started working in homes of wealthy people, caring for their children or doing domestic work.

"You look tired," I said one evening when she walked into our narrow apartment. It was already dark, and she'd put in a long day working two jobs, neither of them well-paying.

She leaned back in the overstuffed chair. "Guess I am," she said as she kicked off her shoes. Her smile caressed me. "What did you learn in school today?" she asked.

No matter how tired she was, if we were still up when she got home, Mother didn't fail to ask about school. As much as anything, her concern for our education began to impress on me that she considered school important.

I was still 8 years old when we moved to Boston, a sometimes serious-minded child who occasionally pondered all the changes that had come into my life. One day I said to myself, "Being 8 is fantastic because when you're 8 you don't have any responsibilities. Everybody takes care of you, and you can just play and have fun."

But I also said, "It's not always going to be this way. So I'm going to enjoy life now."

With the exception of the divorce, the best part of my childhood happened when I was 8 years old. First, I had the most spectacular Christmas of my life. Curtis and I had a wonderful time Christmas shopping, then our aunt and uncle swamped us with toys. Mother too, trying to make up for the loss of our father, bought us more than she ever had before.

One of my favorite gifts was a scale model 1959 Buick with friction wheels. But the chemistry set topped even the toy Buick. Never, before or since, did I have a toy that held my interest like the chemistry set. I spent hours in the bedroom playing with the set, studying the directions, and working one experiment after another. I turned litmus paper blue and red. I mixed chemicals into strange concoctions and watched in fascination when they fizzled, foamed, or turned different colors. When something I'd created filled the whole apartment with the smell of rotten eggs or worse, I'd laugh until my sides ached.

Second, I had my first religious experience when I was 8 years old. We were Seventh-day Adventists, and one Saturday morning Pastor Ford, at the Detroit Burns Avenue church, illustrated his sermon with a story.

A natural storyteller, Pastor Ford told of a missionary doctor husband and wife who were being chased by robbers in a far-off country. They dodged around trees and rocks, always managing to keep just ahead of the bandits. At last, gasping with exhaustion, the couple stopped short at a precipice. They were trapped. Suddenly, right at the edge of the cliff, they saw a small break in the rock—a split just big enough for them to crawl into and hide. Seconds later, when the men reached the edge of the escarpment, they couldn't find the doctor and his wife. To their unbelieving eyes, the couple had just vanished. After screaming and cursing them, the bandits left.

As I listened, the picture became so vivid that I felt as if I were being chased. The pastor wasn't overly dramatic, but I got caught up in an emotional experience, living their plight as if the wicked men were trying to capture me. I visualized myself being pursued. My breath became shallow with the panic and fear and desperation of that couple. At last when the bandits left, I sighed with relief at being safe.

Pastor Ford looked out over the congregation. "The couple were sheltered and protected," he told us. "They were hidden in the cleft of the rock, and God protected them from harm."

The sermon over, we began to sing the "appeal song." That morning the pastor had selected "He Hideth My Soul in the Cleft of the Rock." He built his appeal around the missionary story and explained our need to flee to "the cleft of the rock," to safety found only in Jesus Christ.

"If we place our faith in the Lord," he said as his gaze swept across the faces in the congregation, "we'll always be safe. Safe in Jesus Christ."

As I listened, my imagination pictured how wonderfully God had taken care of those people who wanted to serve Him. Through my imagination and emotions I lived that story with the couple, and I thought, *That's exactly what I should do—get sheltered in the cleft of the rock.*

Although I was only 8, my decision seemed perfectly natural. Other kids my age were getting baptized and joining the church, so when the message and music touched me emotionally, I responded. Following the custom of our denomination, when Pastor Ford asked if anyone wanted to turn to Jesus Christ, Curtis and I both went up to the front of the church. A few weeks later we were both baptized.

I was basically a good kid and hadn't done anything particularly wrong, yet for the first time in my life I knew I needed God's help. During the next four years I tried to follow the teachings I received at church.

That morning set another milestone for me. I decided I wanted to be a doctor, a missionary doctor.

The worship services and our Bible lessons frequently focused on stories about missionary doctors. Each story of medical missionaries traveling through primitive villages in Africa or India intrigued me. Reports came to us of the physical suffering the doctors relieved and how they helped people to lead happier and healthier lives.

"That's what I want to do," I said to my mother as we walked home. "I want to be a doctor. Can I be a doctor, Mother?"

"Bennie," she said, "listen to me." We stopped walking and Mother stared into my eyes. Then laying her hands on my thin shoulders, she said, "If you ask the Lord for something and believe He will do it, then it'll happen."

"I believe I can be a doctor."

"Then, Bennie, you will be a doctor," she said matter-of-factly, and we started to walk on again.

After Mother's words of assurance, I never doubted what I wanted to do with my life.

Like most kids I didn't have any idea of what a person had to do to become a doctor, but I assumed that if I did well in school, I could do it. By the time I turned 13, I wasn't so sure I wanted to be a missionary, but I never deviated from wanting to enter the medical profession.

We moved to Boston in 1959 and stayed until 1961, when Mother moved us back to Detroit, because she was financially on her feet again. Detroit was home for us, and besides, Mother had a goal in mind. Even though it wasn't possible in the beginning, she planned to go back and reclaim the house we'd lived in.

The house, about the size of many garages today, was one of those early prefab post-World War II square boxes. The whole building probably wasn't a thousand square feet, but it was in a nice area where the people kept their lawns clipped and showed pride in where they lived.

"Boys," she told us as the weeks and months passed, "just wait. We're going back to our house on Deacon Street. We may not be able to afford living in it now, but we'll make it. In the meantime, we can still use the rent we get from it." Not a day passed that Mother didn't talk about going home. Determination burned in her eyes, and I never doubted that we would.

Mother moved us into a multifamily dwelling just across the tracks from a section called Delray. It was a smoggy industrial area crisscrossed with train tracks, housing little sweatshops making auto parts. It was what I'd call an upper-lower-class neighborhood.

The three of us lived on the top floor. My mother worked two and three jobs at a time. At one place she cared for children, and at the next she cleaned house. Whatever kind of domestic work anyone needed, Mother said, "I can do it. If I don't know how right now, I learn fast."

Actually there wasn't much else she could do to make a living, because she had no other skills. She gained a lot of commonsense education on these jobs, because she was clever and alert. As she worked, she carefully observed everything around her.

She was especially interested in the people, because most of the time she worked for the wealthy. She'd come home and tell us, "This is what wealthy people do. This is how successful people behave. Here's how they think." She constantly drilled this kind of information into my brother and me.

"Now you boys can do it too," she'd say with a smile, adding, "and you can do it better!"

Strangely enough, Mother started holding those goals in front of me when I wasn't a good student. No, that's not exactly true. I was the worst student in my whole fifth-grade class at Higgins Elementary School.

My first three years in the Detroit public school system had given me a good foundation. When we moved to Boston, I entered the fourth grade, with Curtis two years ahead of me. We transferred to a small private church school, because Mother thought that would give us a better education than the public schools. Unfortunately, it didn't work out that way. Though Curtis and I both made good grades, the work was not as demanding as it could have been, and when we transferred back to the Detroit public school system I had quite a shock.

Higgins Elementary School was predominately White. Classes were tough, and the fifth graders that I joined could outdo me in every single subject. To my amazement, I didn't understand anything that was going on. I had no competition for the bottom of the class. To make it worse, I seriously believed I'd been doing satisfactory work back in Boston.

Being at the bottom of the class hurt enough by itself, but the teasing and taunting from the other kids made me feel worse. As kids will do, there was the inevitable conjecture about grades after we'd taken a test.

Someone invariably said, "I know what Carson got!"

"Yeah! A big zero!" another would shoot back.

"Hey, dummy, think you'll get one right this time?"

"Carson got one right last time. You know why? He was trying to put down the wrong answer."

Sitting stiffly at my desk, I acted as if I didn't hear them. I wanted them to think I didn't care what they said. But I did care. Their words hurt, but I wouldn't allow myself to cry or run away. Sometimes a smile plastered my face when the teasing began. As the weeks passed, I accepted that I was at the bottom of the class because that's where I deserved to be.

I'm just dumb. I had no doubts about that statement, and everybody else knew it too.

Although no one specifically said anything to me

about my being Black, I think my poor record rein-
forced my general impression that Black kids just were
not as smart as White ones. I shrugged, accepting the
reality—that's the way things were supposed to be.

Looking back after all these years, I can almost still
feel the pain. The worst experience of my school life
happened in the fifth grade after a math quiz. As usual,
Mrs. Williamson, our teacher, had us hand our papers to
the person seated behind us for grading while she read
the answers aloud. After grading, each test went back to
its owner. Then the teacher called our names, and we
reported our own grade aloud.

The test contained 30 problems. The girl who
corrected my paper was the ringleader of the kids who
teased me about being dumb.

Mrs. Williamson started calling the names. I sat in
the stuffy classroom, my gaze traveling from the bright
bulletin board to the wall of windows covered with
paper cutouts. The room smelled of chalk and children,
and I ducked my head, dreading to hear my name. It was
inevitable. "Benjamin?" Mrs. Williamson waited for me
to report my score.

I mumbled my reply.

"Nine!" Mrs. Williamson dropped her pen, smiled at
me, and said with real enthusiasm, "Why, Benjamin,
that's wonderful!" (For me to score 9 out of 30 was
incredible.)

Before I realized what was going on, the girl behind
me yelled out, "Not nine!" She snickered. "He got *none.*
He didn't get any of them right." Her snickers were
echoed by laughs and giggles all over the room.

"That's enough!" the teacher said sharply, but it was
too late. The girl's harshness cut out my heart. I don't
think I ever felt so lonely or so stupid in my whole life.
It was bad enough that I missed almost every question
on just about every test, but when the whole class—at
least it seemed like everyone there—laughed at my
stupidity, I wanted to drop through the floor.

Tears burned my eyes, but I refused to cry. I'd die
before I let them know how they hurt me. Instead, I

slapped a don't-care smile on my face and kept my eyes on my desk and the big round zero on the top of my test.

I could easily have decided that life was cruel, that being Black meant everything was stacked against me. And I might have gone that way except for two things that happened during fifth grade to change my perception of the whole world.

4 TWO POSITIVES

I don't know," I said as I shook my head. "I mean, I can't be sure." Again I felt stupid from the top of my head to the bottom of my sneakers. The boy in front of me had read every single letter on the chart down to the bottom line without any trouble. I couldn't see well enough to read beyond the top line.

"That's fine," the nurse said to me, and the next child in line stepped up to the eye-examination chart. Her voice was brisk and efficient. "Remember now, try to read without squinting."

Halfway through my fifth grade the school gave us a compulsory eye examination.

I squinted, tried to focus, and read the first line —barely.

The school provided glasses for me, free. When I went to get fitted, the doctor said, "Son, your vision is so bad you almost qualify to be labeled handicapped."

Apparently my eyes had worsened gradually, and I had no idea they were so bad. I wore my new glasses to school the next day. And I was amazed. For the first time I could actually see the writing on the chalkboard from the back of the classroom. Getting glasses was the first positive thing to start me on my climb upward from the bottom of the class. Immediately after getting my vision corrected my grades improved—not greatly, but at least I was moving in the right direction.

When the mid-term report cards came out, Mrs. Williamson called me aside. "Benjamin," she said, "on the whole you're doing so much better." Her smile of approval made me feel like I could do better yet. I knew she wanted to encourage me to improve.

I had a *D* in math—but that did indicate improvement. At least I hadn't failed.

Seeing that passing grade made me feel good. I thought, *I made a* D *in math. I'm improving. There's hope for me. I'm not the dumbest kid in the school.* When a kid like me who had been at the bottom of the class for the first half of the year suddenly zoomed upward—even if only from *F* to *D*—that experience gave birth to hope. For the first time since entering Higgins School I knew I could do better than some of the students in my class.

Mother wasn't willing to let me settle for such a lowly goal as that! "Oh, it's an improvement all right," she said. "And, Bennie, I'm proud of you for getting a better grade. And why shouldn't you? You're smart, Bennie."

Despite my excitement and sense of hope, my mother wasn't happy. Seeing my improved math grade and hearing what Mrs. Williamson had said to me, she started emphasizing, "But you can't settle for just barely passing. You're too smart to do that. You can make the top math grade in the class."

"But, Mother, I didn't fail," I moaned, thinking she hadn't appreciated how much my work had improved.

"All right, Bennie, you've started improving," Mother said, "and you're going to keep on improving."

"I'm trying," I said. "I'm doing the best I can."

"But you can do still better, and I'm going to help you." Her eyes sparkled. I should have known that she had already started formulating a plan. With Mother, it wasn't enough to say, "Do better." She would find a way to show me how. Her scheme, worked out as we went along, turned out to be the second positive factor.

My mother hadn't said much about my grades until the report cards came out at mid-year. She had believed the grades from the Boston school reflected progress. But once she realized how badly I was doing at Higgins Elementary, she started in on me every day.

However, Mother never asked, "Why can't you be like those smart boys?" Mother had too much sense for that. Besides, I never felt she wanted me to compete with my classmates as much as she wanted me to do my best.

"I've got two smart boys," she'd say. "Two mighty smart boys."

"I'm doing my best," I'd insist. "I've improved in math."

"But you're going to do better, Bennie," she told me one evening. "Now, since you've started getting better in math, you're going to go on, and here's how you'll do it. First thing you're going to do is to memorize your times tables."

"My times tables?" I cried. I couldn't imagine learning so much. "Do you know how many there are? Why that could take a year!"

She stood up a little taller. "I only went through third grade, and I know them all the way through my twelves."

"But, Mother, I can't—"

"You can do it, Bennie. You just have to set your mind to concentrating. You work on them, and tomorrow when I get home from work we'll review them. We'll keep on reviewing the times tables until you know them better than anyone else in your class!"

I argued a little more, but I should have known better.

"Besides"—here came her final shot—"you're not to go outside and play after school tomorrow until you've learned those tables."

I was almost in tears. "Look at all these things!" I

cried, pointing to the columns in the back of my math book. "How can anyone learn all of them?"

Sometimes talking to Mother was like talking to a stone. Her jaw was set, her voice hard. "You can't go outside and play until you learn your times tables."

Mother wasn't home, of course, when school let out, but it didn't occur to me to disobey. She had taught Curtis and me properly, and we did what she told us.

I learned the times tables. I just kept repeating them until they fixed themselves in my brain. Like she promised, that night Mother went over them with me. Her constant interest and unflagging encouragement kept me motivated.

Within days after learning my times tables, math became so much easier that my scores soared. Most of the time my grades reached as high as the other kids in my class. I'll never forget how I felt after another math quiz when I answered Mrs. Williamson with "Twenty-four!"

I practically shouted as I repeated, "I got 24 right."

She smiled back at me in a way that made me know how pleased she was to see my improvement. I didn't tell the other kids what was going on at home or how much the glasses helped. I didn't think most of them cared.

Things changed immediately and made going to school more enjoyable. Nobody laughed or called me the dummy in math anymore! But Mother didn't let me stop with memorizing the times tables. She had proven to me that I could succeed in one thing. So she started the next phase of my self-improvement program to make me come out with the top grades in every class. The goal was fine, I just didn't like her method.

"I've decided you boys are watching too much television," she said one evening, snapping off the set in the middle of a program.

"We don't watch that much," I said. I tried to point out that some programs were educational and that all the kids in my class watched television, even the smartest ones.

As if she didn't hear a word I said, she laid down the law. I didn't like the rule, but her determination to see us improve changed the course of my life. "From now on, you boys can watch no more than three programs a week."

"A week?" Immediately I thought of all the wonderful programs I would have to miss.

Despite our protests, we knew that when she decided we couldn't watch unlimited television, she meant it. She also trusted us, and both of us adhered to the family rules because we were basically good kids.

Curtis, though a bit more rebellious than I was, had done better in his schoolwork. Yet his grades weren't good enough to meet Mother's standards either. Evening after evening Mother talked with Curtis, working with him on his attitude, urging him to want to succeed, pleading with him not to give up on himself. Neither of us had a role model of success, or even a respected male figure to look up to. I think Curtis, being older, was more sensitive to that than I was. But no matter how hard she had to work with him, Mother wouldn't give up. Somehow, through her love, determination, encouragement, and laying down the rules, Curtis became a more reasonable type of person and started to believe in himself.

Mother had already decided how we would spend our free time when we weren't watching television. "You boys are going to go to the library and check out books. You're going to read at least two books every week. At the end of each week you'll give me a report on what you've read."

That rule sounded impossible. Two books? I had never read a whole book in my life, except those they made us read in school. I couldn't believe I could ever finish one whole book in a short week.

But a day or two later found Curtis and me dragging our feet the seven blocks from home to the public library. We grumbled and complained, making the journey seem endless. But Mother had spoken, and it didn't occur to either of us to disobey. The reason? We respected her. We knew she meant business and knew we'd better mind. But, most important, we loved her.

"Bennie," she said again and again, "if you can read, honey, you can learn just about anything you want to know. The doors of the world are open to people who can read. And my boys are going to be successful in life, because they're going to be the best readers in the school."

As I think about it, I'm as convinced today as I was back in the fifth grade, that my mother meant that. She believed in Curtis and me. She had such faith in us, we didn't dare fail! Her unbounded confidence nudged me into starting to believe in myself.

Several of Mother's friends criticized her strictness. I heard one woman ask, "What are you doing to those boys, making them study all the time? They're going to hate you."

"They can hate me," she answered, cutting off the woman's criticism, "but they're going to get a good education just the same!"

Of course I never hated her. I didn't like the pressure, but she managed to make me realize that this hard work was for my good. Almost daily, she'd say, "Bennie, you can do anything you set yourself to do."

Since I've always loved animals, nature, and science, I chose library books on those topics. And while I was a horrible student in the traditionally academic subjects, I excelled in fifth-grade science.

The science teacher, Mr. Jaeck, understood my interest and encouraged me by giving me special projects, such as helping other students identify rocks, animals, or fish. I had the ability to study the markings on a fish, for instance, and from then on I could identify that species. No one else in the class had that knack, so I had my chance to shine.

Initially, I went to the library and checked out books about animals and other nature topics. I became the fifth-grade expert in anything of a scientific nature. By the end of the year I could pick up just about any rock along the railroad tracks and identify it. I read so many fish and water life books, that I started checking streams for insects. Mr. Jaeck had a microscope, and I loved to get water samples to examine the various protozoa under the magnified lenses.

Slowly the realization came that I was getting better in all my school subjects. I began looking forward to my trips to the library. The staff there got to know Curtis and me, offering suggestions on what we might like to read. They would inform us about new books as they came in. I thrived on this new way of life, and soon my interests widened to include books on adventure and scientific discoveries.

By reading so much, my vocabulary automatically improved along with my comprehension. Soon I became the best student in math when we did story problems.

Up until the last few weeks of fifth grade, aside from math quizzes, our weekly spelling bees were the worst part of school for me. I usually went down on the first word. But now, 30 years later, I still remember the word that really got me interested in learning how to spell.

The last week of fifth grade we had a long spelling bee in which Mrs. Williamson made us go through every spelling word we were supposed to have learned that year. As everyone expected, Bobby Farmer won the spelling bee. But to my surprise, the final word he spelled correctly to win was agriculture .

I can spell that word, I thought with excitement. I had learned it just the day before from my library book. As the winner sat down, a thrill swept through me—a yearning to achieve—more powerful than ever before. "I can spell agriculture," I said to myself, "and I'll bet I can learn to spell any other word in the world. I'll bet I could learn to spell better than Bobby."

Learning to spell better than Bobby Farmer really challenged me. Bobby was clearly the smartest boy in the fifth grade. Another kid named Steve Kormos had earned the reputation as being the smartest kid before Bobby Farmer came along. Bobby Farmer impressed me during a history class because the teacher mentioned flax, and none of us knew what she was talking about.

Then Bobby, still new in school, raised his hand and explained to the rest of us about flax—how and where it was grown, and how the women spun the fibers into linen. As I listened, I thought, *Bobby sure knows a lot about flax. He's really smart.* Suddenly, sitting there in the classroom with spring sunshine slanting through the windows, a new thought flashed through my mind. *I can learn about flax or any subject through reading. It is like Mother says—if you can read, you can learn just about anything.* I kept reading all through the summer, and by the time I began sixth grade I had learned to spell a lot of words without conscious memorization. In the sixth grade, Bobby was still the smartest boy in the class, but I was starting to gain ground on him.

After I started pulling ahead in school, the desire to be smart grew stronger and stronger. One day I thought, *It must be a lot of fun for everybody to know you're the smartest kid in the class.* That's the day I decided that the only way to know for sure how that would feel was to become the smartest.

As I continued to read, my spelling, vocabulary, and comprehension improved, and my classes became much more interesting. I improved so much that by the time I entered seventh grade at Wilson Junior High, I was at the top of the class.

But just making it to the top of the class wasn't my real goal. By then, that wasn't good enough for me. That's where Mother's constant influence made the difference. I didn't work hard to compete and to be better than the other kids as much as I wanted to be the very best I could be—for me.

Most of the kids who had gone to school with me in fifth and sixth grade also moved on to Wilson. Yet our relationships had drastically changed during that two-year period. The very kids who once teased me about being a dummy started coming up to me, asking, "Hey, Bennie, how do you solve this problem?"

Obviously I beamed when I gave them the answer. They respected me now because I had earned their respect. It was fun to get good grades, to learn more, to know more than was actually required.

Wilson Junior High was still predominantly White, but both Curtis and I became outstanding students there. It was at Wilson that I first excelled among White kids. Although not a conscious thing on my part, I like to look back and think that my intellectual growth helped to erase the stereotypical idea of Blacks being intellectually inferior.

Again, I have my mother to thank for my attitude. All through my growing up, I never recall hearing her say things such as "White people are just . . ." This uneducated woman, married at 13, had been smart enough to figure out things for herself and to emphasize to Curtis and me that people are people. She never gave vent to racial prejudice and wouldn't let us do it either.

Curtis and I encountered prejudice, and we could have gotten caught up in it, especially in those days —the early 1960s.

Three incidents of racial prejudice directed against us stand out in my memory.

First, when I started going to Wilson Junior High, Curtis and I often hopped a train to get to school. We had fun doing that because the tracks ran parallel to our school route. While we knew we weren't supposed to hop trains, I placated my conscience by deciding to get on only the slower trains.

My brother would grab on to the fast-moving trains which had to slow down at the crossing. I envied Curtis as I watched him in action. When the faster trains came through, just past the crossing he would throw his clarinet on one of the flat cars near the front of the train. Then he'd wait and catch the last flat car. If he didn't get on and make his way to the front, he knew he'd lose his clarinet. Curtis never lost his musical instrument.

We chose a dangerous adventure, and every time we jumped on a train my body tingled with excitement. We not only had to jump and catch a car railing and hold on, but we had to make sure the railroad security men never caught us. They watched for kids and hoboes who hopped the trains at crossroads. They never did catch us.

We stopped hopping trains for an entirely different reason. One day when Curtis wasn't with me, as I ran along the tracks, a group of older boys—all White —came marching toward me, anger written on their faces. One of them carried a big stick.

"Hey, you! Nigger boy!"

I stopped and stared, frightened and silent. I've always been extremely thin and must have looked terribly defenseless—and I was. The boy with the stick whacked me across the shoulder. I recoiled, not sure what would happen next. He and the other boys stood in front of me and called me every dirty name they could think of.

My heart pounded in my ears, and sweat poured down my sides. I looked down at my feet, too scared to answer, too frightened to run.

"You know you nigger kids ain't supposed to be going to Wilson Junior High. If we ever catch you again, we're going to kill you." His pale eyes were cold as death. "You understand that?"

My gaze never left the ground. "Guess so," I muttered.

"I said, 'Do you understand me, nigger boy?' " the big boy prodded.

Fear choked me. I tried to speak louder. "Yes."

"Then you get out of here as fast as you can run. And you'd better be keeping an eye out for us. Next time, we're going to kill you!"

I ran then, as fast as I could, and didn't slow down until I reached the schoolyard. I stopped using that route and went another way. From then on I never hopped another train, and I never saw the gang again.

Certain that my mother would have yanked us out of school right away, I never told her about the incident.

A second, more shocking episode occurred when I was in the eighth grade. At the end of each school year the principal and teachers handed out certificates to the one student who had the highest academic achievement in the seventh, eighth, and ninth grades respectively. I won the certificate in the seventh grade, and that same year Curtis won for the ninth grade. By the end of eighth grade, people had pretty much come to accept the fact that I was a smart kid. I won the certificate again the following year. At the all-school assembly one of the teachers presented my certificate. After handing it to me she remained up in front of the entire studeny body and looked out across the auditorium. "I have a few words I want to say right now," she began, her voice unusually high. Then, to my embarrassment, she bawled out the White kids because they had allowed me to be number one. "You're not trying hard enough," she told them.

While she never quite said it in words, she let them know that a Black person shouldn't be number one in a class where everyone else was White.

As the teacher continued to berate the other students, a number of things tumbled about inside my mind. Of course, I was hurt. I had worked hard to be the top of my class—probably harder than anyone else in the school—and she was putting me down because I wasn't the same color. On the one hand I thought, *What a turkey this woman is!* Then an angry determination welled up inside. *I'll show you and all the others too!*

I couldn't understand why this woman talked the way she did. She had taught me herself in several classes, had seemed to like me, and she clearly knew that I had earned my grades and merited the certificate of achievement. Why would she say all these harsh things? Was she so ignorant that she didn't realize that people are just people? That their skin or their race doesn't make them smarter or dumber? It also occurred to me that, given enough situations, there are bound to be instances where minorities are smarter. Couldn't she realize that?

Despite my hurt and anger, I didn't say anything. I sat quietly while she railed. Several of the White kids glanced over at me occasionally, rolling their eyes to let me understand their disgust. I sensed they were trying to say to me, "What a dummy she is!"

Some of those very kids, who, three years earlier,

had taunted me, had become my friends. They were feeling embarrassed, and I could read resentment on several faces.

I didn't tell Mother about that teacher. I didn't think it would do any good and would only hurt her feelings.

The third incident that stands out in my memory centered around the football team. In our neighborhood we had a football league. When I was in the seventh grade, playing football was the big thing in athletics.

Naturally, both Curtis and I wanted to play. Neither of us Carsons were large to begin with. In fact, compared to the other players, we were quite small. But we had one advantage. We were fast—so fast that we could outrun everybody else on the field. Because the Carson brothers made such good showings, our performance apparently upset a few of the White people.

One afternoon when Curtis and I left the field after practice, a group of White men, none of them over 30 years old, surrounded us. Their menacing anger showed clearly before they said a word. I wasn't sure if they were part of the gang that had threatened me at the railroad crossing. I only knew I was scared.

Then one man stepped forward. "If you guys come back we're going to throw you into the river," he said. Then they turned and walked away from us.

Would they have carried out their threat? Curtis and I weren't as concerned about that as we were with the fact that they didn't want us in the league.

As we walked home, I said to my brother, "Who wants to play football when your own supporters are against you?"

"I think we can find better things to do with our time," Curtis said.

We never said anything to anyone about quitting, but we never went back to practice. Nobody in the neighborhood ever asked us why. To Mother I said, "We decided not to play football." Curtis said something about studying more.

We had decided to say nothing to Mother about the threat, knowing that if we did, she'd be worried sick about us. As an adult looking backward, it's ironic about our family. When we were younger, through her silence Mother had protected us from the truth about Dad and her emotional problems. Now it was our turn to protect her so she wouldn't worry. We chose the same method.

5 A BOY'S BIG PROBLEM

Know what the Indians did with General Custer's worn-out clothes?" the gang leader asked.

"Tell us," one of his cohorts shot back with exaggerated interest.

"They saved them and now our man Carson wears them!"

Another kid nodded vigorously. "Sure looks it."

I could feel the heat rising up my neck and cheeks. The guys were at it again.

"Get close enough and you'll believe it," the first fellow laughed, " 'cuz they smell like they're a hundred years old!"

New in the grade of 8-A at Hunter Junior High, I found capping an embarrassing and painful experience. The term comes from the word capitalize and is slang that means to get the better of another person. The idea was to make the most sarcastic remark possible, throwing in a quick barb to keep it humorous. Capping was always done within earshot of the victim, and the best targets were the kids whose clothes were a little out of style. The best cappers waited until a group collected around the violator. Then they'd compete to see who could say the funniest and most insulting things.

I was a special target. For one thing, clothes hadn't meant much to me then, and they don't today. Except for a brief period in my life, I've not been much concerned about what I wore, because like Mother always said, "Bennie, what's inside counts the most. Anybody can dress up on the outside and be dead inside."

I hated leaving Wilson Junior High at the middle of the eighth grade but was excited to be moving back to our old house. As I said to myself, "We're going home again!" That was the most important thing of all.

Because of my mother's frugality, our financial situation had gradually improved. Mother was finally able to get enough money, and we moved back to the house where we lived before my parents divorced.

Despite the smallness of the house, it was home. Today I see it more realistically—more like a matchbox. But to the three of us then, the house seemed like a mansion, a really fabulous place.

But moving home meant the need to change schools. While Curtis went on to Southwestern High School, I enrolled in Hunter Junior High, a predominantly Black school with about 30 percent of the students White.

Classmates immediately recognized me as a smart kid. Although I wasn't quite at the top, only one or two others passed me in grades. I had grown used to academic success, enjoyed it, and decided to stay on top.

At that point, however, I felt a new pressure—one that I hadn't been subjected to before. Besides the capping, I faced the constant temptation to become one of the guys. I'd never had to be involved in this kind of thing before in order to be accepted. In the other schools, kids looked up to me because of my top grades. But at Hunter Junior High, academics came a little farther down the line.

Being accepted by the in-group meant wearing the right clothes, going to the places where the guys hung out, and playing basketball. Even more important, to be part of the in-group, kids had to learn to cap on others.

I couldn't ask my mother to buy me the kind of clothes that would put me on their social-acceptance level. While I may not have understood how hard my mother worked, I knew she was trying to keep us off of public assistance. By the time I went into ninth grade, Mother had made such strides that she received nothing except food stamps. She couldn't have provided for us and kept up the house without that subsidy.

Because she wanted to do the best she could for Curtis and me, she skimped on herself. Her clothes were clean and respectable, but they weren't stylish. Of course, being a kid, I never noticed, and she never complained.

For the first few weeks I didn't say anything when the guys capped on me. My lack of response only encouraged them to bear down, and they capped on me mercilessly. I felt horrible, left out, and hurt because I didn't fit in. Walking home alone, I'd wonder, *What's wrong with me? Why can't I belong? Why do I have to be different?* I comforted myself by saying, "They're just a bunch of buffoons. If this is how they get their enjoyment, they can go ahead, but I'm not going to play their silly game. I'm going to be successful, and one day I'll show all of them."

Despite my defensive words, I still felt left out and rejected. And, like most people, I wanted to belong and didn't like being an outsider. Unfortunately, after a while their attitude rubbed off on me until eventually the disease infected me too. Then I said to myself, "All right, if you guys want to cap, I'll show you how to cap."

The next day I waited for the capping to start. And it did. A ninth grader said, "Man, that shirt you're wearing has been through World War I, World War II, World War III, and World War IV."

"Yeah," I said, "and your mama wore it."

Everybody laughed.

He stared at me, hardly believing what I'd said. Then he started to laugh too. He slapped me on the back. "Hey, man, that's OK."

My esteem rose right then. Soon I capped on the top cappers throughout the whole school. It felt great to be recognized for my sharp tongue.

From then on when anyone capped on me, I'd turn it around and fling it into their faces—which was the idea of the game. Within weeks the in-crowd stopped tormenting me. They didn't dare direct any sarcasm my way because they knew I would come up with something better.

Once in a while, students ducked out of the way when they saw me coming. I didn't let them get away even then. "Hey, Miller! I'd hide my face too if I looked that ugly!"

A mean remark? Certainly, but I comforted myself by saying, "Everybody does it. Outcapping everyone else is the only way to survive." Or sometimes I'd say, "He knows I didn't really mean it."

It didn't take long for me to forget how it felt to be the object of capping. My taking over the game solved one great problem for me.

Unfortunately, it didn't solve what to do about clothes.

Aside from being ostracized for my clothes, the kids called me poor a lot. And to their thinking, if you were poor, you were no good. Oddly enough, none of the students were well-off and had no right to talk about anybody else. But as a young teenager, I didn't reason that out. I felt the stigma of being poor most acutely because I didn't have a father. Most of the kids I knew had two parents, and that convinced me that they were better off.

During ninth grade one task brought more embarrassment to me than anything else. As I've said, we received food stamps and couldn't have made it without them.

Occasionally my mother sent me to the store to buy bread or milk with the stamps. I hated to go, fearing one of my friends would see what I was doing. If anyone I knew came up to the checkout counter, I'd pretend that I had forgotten something and duck down one of the aisles until he left. Waiting until nobody else stood in line, I'd rush forward with the items I had to buy.

I could accept being poor, but I died a thousand deaths thinking that other kids would know it. If I had thought more logically about the food stamps, I would have realized that quite a few of my friends' families used them too. Yet every time I left the house with the stamps burning in my pocket, I worried that someone might see me or hear about my using food stamps and then talk about me. So far as I know, no one ever did.

The ninth grade stands out as a pivotal time in my life. As an *A* student I could stand up intellectually with the best. And I could hold my own with the best—or worst—of my classmates. It was a time of transition. I was leaving childhood and beginning to think seriously about the future and especially about my desire to be a doctor.

By the time I hit the tenth grade, however, the peer pressure had gotten to be too much for me. Clothes were my biggest problem. "I can't wear these pants," I'd tell Mother. "Everyone will laugh at me."

"Only stupid people laugh at what you wear, Bennie," she'd say. Or, "It's not what you're wearing that makes the difference."

"But, Mother," I'd plead. "Everybody I know has better clothes than I do."

"Maybe so," she'd patiently tell me. "I know a lot of people who dress better than I do, but that doesn't make them better."

Just about every day, I begged and pressured my mother, insisting that I had to have the right kind of clothes. I knew exactly what I meant by the right kind: Italian knit shirts with suede fronts, silk pants, thick-and-thin silk socks, alligator shoes, stingy brim hats, leather jackets, and suede coats. I talked about those clothes constantly, and it seemed like I couldn't think about anything else. I had to have those clothes. I had to be like the in-crowd.

Mother was disappointed in me and I knew it, but all I could think of was my poor wardrobe and my need for acceptance. Instead of coming directly home after school and doing my homework, I played basketball. Sometimes I stayed out until ten o'clock, and a few times until eleven. When I came home I knew what to expect, and I prepared myself to endure it.

"Bennie, can't you see what you're doing to yourself? It's more than just disappointing me. You're going to ruin your life staying out all hours and begging for nothing but fine clothes."

"I'm not ruining my life," I insisted, because I didn't want to listen. I couldn't have heard anything because my immature mind focused on being like everybody else.

"I've been proud of you, Bennie," she would say. "You've worked hard. Don't lose all of that now."

"I'll keep on doing all right," I'd snap back. "I'll be OK. Haven't I been bringing home good grades?"

She couldn't argue with me on that issue, but I know she worried. "All right, son," she finally told me.

Then, after weeks of my pleading for new clothes, Mother said the words I wanted to hear. "I'll try to get some of those fancy clothes for you. If that's what it takes to make you happy, you'll have them."

"They'll make me happy," I said. "They will."

It's hard for me to believe how insensitive I was back then. Without thinking about her needs, I let Mother go without to buy me clothes that would help me dress like the in-crowd. But I never had enough. Now I realize that no matter how many Italian shirts, leather jackets, or alligator shoes she bought, they would never have been enough.

My grades dropped. I went from the top of the class to being a *C* student. Even worse, achieving only average grades didn't bother me because I was part of the in-group. I hung out with the popular guys. They invited me to their parties and jam sessions. And fun—I was having more fun than I'd ever had in my life because I was one of the guys.

I just wasn't very happy.

I had strayed from the important and basic values in my life. To explain that statement, I have to go back to my mother again and tell you about a visit from Mary Thomas.

When my mother was in the hospital to deliver me, she had her first contact with Seventh-day Adventists. Mary Thomas was visiting in the hospital and started talking to her about Jesus Christ. Mother listened politely but had little interest in what she had to say.

Later, as I've already mentioned, Mother was so emotionally hurt that she checked herself into a mental hospital. At one point, she seriously considered committing suicide by saving up her daily medication and taking all the pills at once. Then one afternoon a woman visited my mother in the hospital. She had met the woman once before—Mary Thomas.

This quiet but zealous woman began talking to her about God. That in itself was nothing new. From the time she was a little girl in Tennessee, Mother had heard about God. Yet Mary Thomas approached religion differently. She didn't try to force anything on Mother or tell her how sinful she was. Instead, Mary Thomas simply expressed her own beliefs and paused occasionally to read verses from the Bible that explained the basis for her faith.

More important than her teaching, Mary genuinely cared about Mother. And right then Mother needed someone to care.

Even before the divorce, Mother was a desperate woman with two young kids and no idea how to take care of them if things didn't work out. She was ostracized by many who felt she was unconventional. Then along came Mary Thomas with what seemed like a single ray of hope. "There is another source of strength, Sonya," the visitor said. "And this strength can be yours."

Those were exactly the words she needed as a stabilizing force in her life. Mother finally understood that she wasn't all alone in the world.

Over a period of weeks, Mary went over the teachings of her church, and Mother slowly came to believe in a loving God who expresses that love through Jesus Christ.

Day after day Mary Thomas talked patiently with Mother, answering questions, and listening to anything she wanted to say.

Mother's third-grade education prevented her from reading most of the Bible passages, but her visitor didn't give up. She stayed at it, reading everything aloud. And through that woman's influence my mother began to study and read for herself.

Even though Mother could barely read, once she decided to learn, through hours of practice she taught herself to read well. Mother started to read the Bible, often sounding out the words, sometimes still not understanding; but she persisted. That was her determination at work. Eventually she was able to read relatively sophisticated material.

Aunt Jean and Uncle William, with whom we stayed after my parents' divorce, had become Adventists in Boston. With their encouragement, it wasn't long until Mother grew stronger in her beliefs. Never one to go into anything half-heartedly, she immediately became active and has remained a devout church member. And from the time of her own conversion, she started taking Curtis and me to church with her. The Adventist denomination is the only spiritual home I've ever known.

When I was 12 and more mature, I realized that although I'd been emotionally touched at age 8 and even had been baptized, I hadn't understood exactly what being a Christian meant.

By the time I was 12, we had moved and were attending the Sharon Seventh-day Adventist Church in Inkster. After days of thinking about the matter, I spoke with Pastor Smith. "Although I've been baptized," I said,

"I didn't really grasp the significance of what I was doing."

"You do understand now?"

"Oh, yes, I'm 12 now," I said, "and I believe in Jesus Christ. After all, Jesus was 12 when His parents first took Him to the temple in Jerusalem. So I'd like to be baptized again, because I understand and I'm ready now."

Pastor Smith listened sympathically, and having no problem with my request, he rebaptized me.

Yet in looking back, I'm not sure when I actually turned to God. Or perhaps it happened so gradually that I had no awareness of the progression. I do know that when I was 14, I finally understood how God can change us.

It was at age 14 that I confronted the most severe personal problem of my life, one that almost ruined me forever.

6 A TERRIBLE TEMPER

That sure was a dumb thing to say," Jerry taunted as we walked down the hall together after English class. Kids crowded us on all sides, and Jerry's voice rose above the din.

I shrugged. "Guess so." My wrong answer in seventh-grade English had been embarrassing enough. I didn't want to be reminded.

"You guess?" Jerry's laugh was shrill. "Listen, Carson, that was one of the all-time stupid things of the year!"

I turned my eyes toward him. He was taller and heavier, not even one of my close friends. "You've said some pretty dumb things too," I said softly.

"Oh yeah?"

"Yeah. Just last week you—"

Our words flew back and forth, my voice remaining calm while his grew louder and louder. Finally I turned to my locker. I'd just ignore him, and maybe he'd shut up and go away.

My fingers twirled the combination lock. Then, just as I lifted the lock, Jerry shoved me. I stumbled, and my temper flared. I forgot the 20 pounds of muscle he had on me. I didn't see the kids and teachers milling in the hall. I swung at him, lock in hand. The blow slammed into his forehead, and he groaned, staggering backward, blood seeping from a three-inch gash.

Dazed, Jerry slowly lifted his hand to his forehead. He felt the sticky blood and carefully lowered his hand in front of his eyes. He screamed.

Of course the principal called me in. I'd calmed down by then and apologized profusely. "It was almost an accident," I told him. "I never would have hit him if I'd remembered the lock in my hand." I meant it too. I was ashamed. Christians didn't lose their temper like that. I apologized to Jerry and the incident was closed.

And my temper? I forgot about it. I wasn't the kind of guy who'd split open a kid's head on purpose.

Some weeks later Mother brought home a new pair of pants for me. I took one look at them and shook my head. "No way, Mother. I'm not going to wear them. They're the wrong kind."

"What do you mean 'wrong kind'?" she countered. She was tired. Her voice firm. "You need new pants. Now just wear these!"

I flung them back at her. "No," I yelled. "I'm not going to wear these ugly things."

She folded the pants across the back of the plastic kitchen chair. "I can't take them back." Her voice was patient. "They were on special."

"I don't care." I spun to face her. "I hate them, and I wouldn't be caught dead in them."

"I paid good money for these pants."

"They're not what I want."

She took a step forward. "Listen, Bennie. We don't always get what we want out of life."

Heat poured through my body, inflaming my face, energizing my muscles. *"I* will!" I yelled. "Just wait and see. I will. I'll—"

My right arm drew back, my hand swung forward. Curtis jumped me from behind, wrestling me away from Mother, pinning my arms to my side.

The fact that I almost hit my mother should have

made me realize how deadly my temper had become. Maybe I knew it but wouldn't admit the truth to myself. I had what I only can label a pathological temper—a disease—and this sickness controlled me, making me totally irrational.

In general I was a good kid. It usually took a lot to make me mad. But once I reached the boiling point, I lost all rational control. Totally without thinking, when my anger was aroused, I grabbed the nearest brick, rock, or stick to bash someone. It was as if I had no conscious will in the matter.

Friends who didn't know me as a kid think I'm exaggerating when I say I had a bad temper. But it's no exaggeration and to make it clear, here are just two more of my crazed experiences.

I can't remember how this one started, but a neighborhood kid hit me with a rock. It didn't hurt, but again, out of that insane kind of anger, I raced to the side of the road, picked up a big rock, and hurled it at his face. I seldom missed when I threw anything. The rock broke his glasses and smashed his nose.

I was in the ninth grade when the unthinkable happened. I lost control and tried to knife a friend. Bob and I were listening to a transistor radio when he flipped the dial to another station. "You call that music?" he demanded.

"It's better than what you like!" I yelled back, grabbing for the dial.

"Come on, Carson. You always—"

In that instant blind anger—pathological anger—took possession of me. Grabbing the camping knife I carried in my back pocket, I snapped it open and lunged for the boy who had been my friend. With all the power of my young muscles, I thrust the knife toward his belly. The knife hit his big, heavy ROTC buckle with such force that the blade snapped and dropped to the ground.

I stared at the broken blade and went weak. *I had almost killed him. I had almost killed my friend.* If the buckle hadn't protected him, Bob would have been lying at my feet, dying or severely wounded. He didn't say anything, just looked at me, unbelieving. "I—I'm sorry," I muttered, dropping the handle. I couldn't look him in the eye. Without a word, I turned and ran home.

Thankfully the house was empty, for I couldn't bear to see anyone. I raced to the bathroom where I could be alone, and locked the door. Then I sank down on the edge of the tub, my long legs stretching across the linoleum, bumping against the sink.

I tried to kill Bob. I tried to kill my friend. No matter how tightly I squeezed my eyes shut, I couldn't escape the image—my hand, my knife, the belt buckle, the broken knife. And Bob's face.

"This is crazy," I finally mumbled. "I must be crazy. Sane people don't try to kill their friends." The rim of the tub felt cool under my hands. I put my hands on my hot face. "I'm doing so well at school, and then I do this."

I'd dreamed of being a doctor since I was 8 years old. But how could I fulfill the dream with such a terrible temper? When angry, I went out of control and had no idea how to stop. I'd never make anything of myself if I didn't control my temper. If only I could do something about the rage that burned inside me.

Two hours passed. The green and brown squiggly snakelike design on the linoleum swam before my eyes. I felt sick to my stomach, disgusted with myself, and ashamed. "Unless I get rid of this temper," I said aloud, "I'm not going to make it. If Bob hadn't worn that big buckle he'd probably be dead, and I'd be on my way to jail or reform school."

Misery washed over me. My sweaty shirt stuck to my back. Sweat trickled down my armpits and my sides. I hated myself, but I couldn't help myself, and so I hated myself even more.

From somewhere deep inside my mind came a strong impression. Pray. My mother had taught me to pray. My teachers at the religious school in Boston often told us that God would help us if we only asked Him. For weeks, for months, I had been trying to control my temper, figuring I could handle it myself. Now, in that small hot bathroom I knew the truth. I could not handle my temper alone.

I felt as though I could never face anyone again. How could I look my mother in the eye? Would she know? How could I ever see Bob again? How could he help but hate me? How could he ever trust me again?

"Lord," I whispered, "You have to take this temper from me. If You don't, I'll never be free from it. I'll end up doing things a lot worse than trying to stab one of my best friends."

Already heavy into psychology (I had been reading *Psychology Today* for a year), I knew that temper was a personality trait. Standard thinking in the field pointed out the difficulty, if not the impossibility, of modifying personality traits. Even today some experts believe that the best we can do is accept our limitations and adjust to them.

Tears streamed between my fingers. "Lord, despite what all the experts tell me, You can change me. You can free me forever from this destructive personality trait."

I wiped my nose on a piece of toilet paper and let it drop to the floor. "You've promised that if we come to You and ask something in faith, that You'll do it. I believe that You can change this in me." I stood up, looking at the narrow window, still pleading for God's help. I couldn't go on hating myself forever for all the terrible things I'd done.

I sank down on the toilet, sharp mental pictures of other temper fits filling my mind. I saw my anger, clenched my fists against my rage. I wouldn't be any good for anything if I couldn't change. *My poor mother,* I thought. *She believes in me. Not even she knows how bad I am.*

Misery engulfed me in darkness. "If you don't do this for me, God, I've got no place else to go."

At one point I'd slipped out of the bathroom long enough to grab a Bible. Now I opened it and began to read in Proverbs. Immediately I saw a string of verses about angry people and how they get themselves into trouble. Proverbs 16:32 impressed me the most: "He who is slow to anger is better than the mighty, and he who rules his spirit than he who takes a city" (RSV).

My lips moved wordlessly as I continued to read. I felt as though the verses had been written just to me, for me. The words of Proverbs condemned me, but they also gave me hope. After a while peace begin to fill my mind. My hands stopped shaking. The tears stopped. During those hours alone in the bathroom, something happened to me. God heard my deep cries of anguish. A feeling of lightness flowed over me, and I knew a change of heart had taken place. I felt different. I was different.

At last I stood up, placed the Bible on the edge of the tub, and went to the sink. I washed my face and hands, straightened my clothes. I walked out of the bathroom a changed young man. "My temper will never control me again," I told myself. "Never again. I'm free."

And since that day, since those long hours wrestling with myself and crying to God for help, I have never had a problem with my temper.

That same afternoon I decided I would read the Bible every day. I've kept that practice as a daily habit and especially enjoy the book of Proverbs. Even now, whenever possible, I pick up my Bible and read the first thing every morning.

The miracle that took place was incredible when I stop to think about it. Some of my psychologically oriented friends insist that I still have the potential for anger. Maybe they're right, but I've lived more than twenty years since that experience, and I've never had another flare-up or even had a serious problem of needing to control my temper.

I can tolerate amazing amounts of stress and ridicule. By God's grace, it still doesn't require any effort to shake off unpleasant, irritating things. God has helped me to conquer my terrible temper, once and forever.

During those hours in the bathroom I also came to realize that if people could make me angry they could control me. Why should I give someone else such power over my life?

Over the years I've chuckled at people who deliberately did things they thought would make me angry. I'm no better than anyone else, but I laugh inside at how foolish people can be, trying to make me angry. They don't have any control over me.

And this is the reason. From that terrible day when I was 14 years old, my faith in God has been intensely personal and an important part of who I am. About that time I started to hum or sing a hymn that has continued to be my favorite, "Jesus Is All the World to Me." Whenever anything irritates me, that hymn dissolves my negativity. I've explained it this way to young people, "I have sunshine in my heart regardless of conditions around me."

I'm not afraid of anything as long as I think of Jesus Christ and my relationship to Him and remember that the One who created the universe can do anything. I also have evidence—my own experience—that God can do anything, because He changed me.

From age 14, I began to focus on the future. My mother's lessons—and those of several of my teachers—were at last paying off.

7 ROTC TRIUMPH

I was 10 years old when I first became interested in Johns Hopkins University Hospital. Back in those days it seemed that every television or newspaper medical story involved somebody at Johns Hopkins. So I said, "That's where I want to go when I become a doctor. Those guys are finding cures and new ways to help sick people."

Although I had no question about wanting to be a doctor, the particular field of medicine wasn't always so clear. For instance, when I was 13 my focus changed from being a general practitioner to becoming a psychiatrist. Watching TV programs featuring psychiatrists convinced me, for they came across as dynamic intellectuals who knew everything about solving anybody's problems. At that same age I was very aware of money and figured that with so many crazy people living in the United States, psychiatrists must make a good living.

If I had any doubts about my chosen career they dissolved after my thirteenth birthday when Curtis gave me a subscription to *Psychology Today*. It was the perfect gift. Not only a great brother but a good friend,

Curtis must have really sacrificed to spend his hard-earned money for me. He was only 15, and his after-school job in the science lab didn't pay a lot.

Curtis was generous but also sensitive to me. Because he knew I was getting interested in psychology and psychiatry, he chose that way to help me. Though I found *Psychology Today* tough reading for a kid my age, I grasped enough from the different articles that I could hardly wait for each issue to arrive. I also read books in that field. For awhile I fancied myself as some sort of local shrink. Other kids came to me with their problems. I was a good listener, and I learned certain techniques for helping others. I'd ask questions like, "Do you want to talk about it?" or "What's troubling you today?"

The kids opened up. Maybe they just wanted a chance to talk about their problems. Some of them were willing to listen. I felt honored to have their confidence and to know that they were willing to tell me their troubles.

"Well, Benjamin," I said to myself one day, "you've found your chosen field, and you're already moving into it."

Not until my days in medical school would that focus shift once more.

In the second half of tenth grade I joined the ROTC. I'll confess that I did that largely because of Curtis. I really admired my brother, although I would never have told him so. Whether he knew it or not, he provided a role model for me. He was one of the people I wanted to emulate. It made me proud to see him in his uniform, his chest plastered with more medals and ribbons than anybody I knew.

My joining the ROTC started another change in my life, helping me to get back on the right track. My brother, then a senior, had reached the rank of captain and was the company commander when I became a private.

Curtis never got caught up in the peer thing and the demand for clothes like I did. He stayed on the honor role and remained a good student right through high school. He graduated near the top of his class and went on to the University of Michigan, eventually majoring in engineering. [1]

After I joined ROTC, another significant person came into my life—a student named Sharper. He had reached the highest rank given to a student—that of a full colonel. Sharper seemed so mature, so self-assured, and yet likeable. *He's incredible,* I thought as I watched him drill the entire ROTC unit. Then came the next thought. *If Sharper could make colonel, why can't I?* At that moment I decided I wanted to be a student colonel.

Because I joined ROTC late (in the second half of tenth grade instead of the beginning of the year like the others), it meant I'd be in ROTC only five semesters instead of six. From the beginning I realized that my chances of ever making it to the top weren't very good, but instead of discouraging me, the thought challenged me. I determined that I would go as far as I possibly could in ROTC before I graduated.

My mother continued to talk to me about my attitude and began to make an impression. She didn't lecture because she was discovering more subtle ways to encourage me. She memorized poems and famous sayings and kept quoting them to me.

Thinking about it now, Mother was incredible, memorizing long poems like Robert Frost's "The Road Not Taken." She often quoted to me a poem called "You Have Yourself to Blame"—a poem I've never been able to find in print. But it's about people offering excuses for failing to do their best. The bottom line was that we have only ourselves to blame. We create our own destiny by the way we do things. We have to take advantage of opportunities and be responsible for our choices.

Mother stayed on me until I fully grasped that I am the one ultimately responsible for my life. I had to take charge if I wanted to amount to anything. Soon my grades zoomed upward again. During both the eleventh and twelfth grades I ranked among the *A* students again. I had gotten back on the right track.

Another influential person in my life was an English

teacher named Mrs. Miller. She took a personal interest in me in ninth-grade English and taught me a lot of extra things after class. She was proud of me because I was such a good student, and she taught me to appreciate good literature and poetry. We'd go over everything I'd done in class that wasn't perfect, and she stayed with me until I corrected every mistake.

In the tenth grade when my grades dropped, she was disappointed. Even though I no longer had her for a teacher, she kept up with me and knew that my indifference to schoolwork caused my grades to fall, because I was just hanging out instead of trying. I felt bad about that, because she was so disappointed. At that point I felt more guilty about disappointing her than I did my mother.

Finally I began to realize that I had myself—and only myself—to blame. The in-group had no power over me unless I chose to give it to them. I started pulling away from them. The clothes issue largely resolved itself because in ROTC we had to wear a uniform three days a week. That meant I had to wear regular clothes only two days a week, and I had enough of the "right" clothes that kids didn't talk about me.

With my clothes problem solved and my changed attitude, once again I started doing very well in school.

Several teachers played important roles in my life during my high school years. They gave me personal attention, encouraged me, and all of them tried to inspire me to keep trying.

I particularly admired and appreciated two men teachers. First, Frank McCotter, the biology teacher. He was White, about five feet nine, medium build, and wore glasses. If I'd first seen him on the street without knowing anything about him, I would have said, "That's a biology teacher."

Mr. McCotter had so much confidence in my abilities that he pushed me to take more responsibility, and he provided me with extra tutoring in the biological sciences. McCotter assigned me the responsibility to design experiments for the other students, to set them up, and to keep the lab running smoothly.

The second teacher, Lemuel Doakes, directed the band. He was Black, well-built, and serious most of the time, although he had a fine sense of humor. Mr. Doakes always demanded perfection. He wouldn't settle for our getting the music right—we had to play it perfectly.

More than being a teacher with interests limited primarily to music, Mr. Doakes encouraged my academic pursuits. He saw that I had musical talent, but he told me, "Carson, you have to put academics first. Always put first things first." I thought that was an admirable attitude for a music teacher.

As much as for his music, I also admired Mr. Doakes for being courageous. He was one of the few teachers who would stand up to the bullies in the school and not let them scare him. He wouldn't tolerate any foolishness. A few students challenged him, but they ended up backing down.

I earned a lot of medals in ROTC for being a member of the rifle team and drill team. I won academic awards and just about every competition offered. Along with this, I received rapid promotion.

One of the big challenges came when I was a master sergeant. Sgt. Bandy, an instructor in the United States Army and head of the ROTC unit at our high school, put me in charge of the fifth-hour ROTC unit because the students were so rambunctious that none of the other student-sergeants could handle them.

"Carson, I'm going to put you in charge of this class," he said. "If you can make anything out of them, I'll promote you to second lieutenant." That was exactly the challenge I needed.

I did two things. First, I tried to get to know the guys in the class and discover what really interested them. Then I structured the classes and the exercises accordingly. I offered extra practice on fancy drill routine at the end of each successful teaching session, and the guys loved doing that.

Second, reverting to my earlier skill at capping on

people paid off. They soon shaped up because, when they didn't do things appropriately, they learned I could make them look bad by capping on them. This method didn't employ the best psychology, but it worked, and they fell into line.

It was just before summer, and I'd been working hard with the class for several weeks when Sgt. Bandy called me into his office. "Carson," he said, "the fifth-hour class is the best unit in the school. You have done a fine job."

And, true to his word, Bandy promoted me to second lieutenant at the end of the year—unheard of in our school. [2]

The promotion allowed me to try for field grade, because only after making second lieutenant could anyone sit for field-grade examinations. The normal route went from second lieutenant to first lieutenant to captain and then to major. After that, few students went on to become lieutenant colonel, and only three in the whole city of Detroit made full colonel.

Sgt. Bandy set it up for me to go up for the field-grade examination. I did so well that he scheduled me to appear before a board of majors and captains in the real Army.

About that time Sgt. Hunt became the first Black sergeant in charge of our ROTC unit, replacing Sgt. Bandy. Sgt. Hunt recognized my leadership ability and, because I was doing so well academically, he took a special interest in me. He'd often take me aside and say things like, "Carson, I've got big plans for you."

Sgt. Hunt used to give me a lot of extra hints and suggestions, sharing his own insights into things that the examiners would want me to know. "Carson," he'd bark, "you gotta learn this and gotta learn it perfect."

I memorized all of the required material. The regular Army officers who conducted the examination asked every possible question from our training manuals —questions about terrain, battle strategies, various weapons, and weapon systems. And I was ready!

When I went up for the field-grade examination, along with representatives from each of the 22 schools in the city, I made the highest score. In fact, my total was (at least then) the highest any student had ever achieved.

To my delighted surprise, I received another promotion—all the way from second lieutenant to lieutenant colonel, again a feat totally unheard of. Naturally, I was elated. Even more of a wonder, this took place during the first part of twelfth grade. I could hardly believe it myself. From the second half of tenth grade (10A) I had gone from private to lieutenant colonel by the time I reached 12B. I still had a full semester of school left, and another field-grade examination was coming up. That meant I actually had an opportunity to become colonel. If I made it, I would be one of three ROTC colonels in Detroit.

I sat for the exam again and did the best of all the competitors. I was made city executive officer over all the schools.

I had realized my dream. I had gotten all the way to colonel even though I had joined ROTC late. Several times I thought, *Well, Curtis, you got me started, and you made captain. I've passed you, but I wouldn't have gotten into the ROTC if you hadn't done it first.*

At the end of my twelfth grade I marched at the head of the Memorial Day parade. I felt so proud, my chest bursting with ribbons and braids of every kind. To make it more wonderful, we had important visitors that day. Two soldiers who had won the Congressional Medal of Honor in Viet Nam were present. More exciting to me, General William Westmoreland (very prominent in the Viet Nam war) attended with an impressive entourage. Afterward, Sgt. Hunt introduced me to General Westmoreland, and I had dinner with him and the Congressional Medal winners. Later I was offered a full scholarship to West Point.

I didn't refuse the scholarship outright, but I let them know that a military career wasn't where I saw myself going. As overjoyed as I felt to be offered such a scholarship, I wasn't really tempted. The scholarship would have obligated me to spend four years in military service after I finished college, precluding my chances to go on to medical school. I knew my direction—I wanted to be a doctor, and nothing would divert me or stand in the way.

Of course the offer of a full scholarship flattered me. I was developing confidence in my abilities—just like my mother had been telling me for at least the past ten years. Unfortunately I carried it a little too far. I started to believe that I was one of the most spectacular and smartest people in the world. After all, I had made this unprecedented showing in ROTC, and I stood at the top of my school academically. The big colleges wrote to me and sent out their representatives to recruit me.

Meeting representatives from places like Harvard and Yale made me feel special and important because they wanted to recruit me. Few of us get enough experience at feeling special and important, and I was no exception. I didn't know how to handle all the attention. The school reps flocked around me because of my high academic achievements, and because I had done exceptionally well on the Scholastic Aptitude Test (SAT), ranking somewhere in the low ninetieth percentile —again, unheard of from a student in the inner city of Detroit.

I laugh sometimes when I think of my secret for scoring so high on the SAT. Back when my mother would allow us to watch only two or three television shows and insisted that we read two books a week, I did just that. One program—my favorite—was the General Electric *College Bowl.* On that program—a quiz show— students from colleges around the country sat as contestants and competed with each other. The master of ceremonies asked factual questions and challenged the knowledge of those students.

All week I looked forward to Sunday nights. In my mind, I had already focused on another secret goal—to be a contestant on the program. To get the chance to appear, I knew I'd have to be knowledgeable in many subjects, so I broadened my range of reading interests. Having inherited a job in the science laboratory after

Curtis graduated helped me tremendously because the science teachers saw my desire to know more. They gave me extra tutoring and suggested books or articles for me to read. Although I was doing well in most of the academic subjects, I realized I didn't know a lot about the arts.

I started going downtown after school to the Detroit Institute of Arts. I walked through the exhibit rooms until I knew all the paintings in the main galleries. I checked out library books about various artists and was really taking in all of that material. Before long I could recognize the masters' paintings, name the works themselves, cite the artists' names and their styles. I learned all kinds of information, such as when the artists lived and where they received their training. I soon could recognize the paintings or artists like a flash when questions came up about them on *College Bowl.*

Next, I had to learn about classical music if I wanted to compete. When I started that phase, I used to receive weird looks from people. For instance, I'd be out on the lawn digging up weeds or trimming the grass and have my portable radio playing classical music. That was considered strange behavior for a Black kid in Motown. Everybody else was listening to jam and bebop.

In truth, I didn't much like the classical music. But here again, Curtis played a decisive role in my life. By then he was in the Navy, and once when he came home on leave he brought a couple of records. One of them was Schubert's *Eighth Symphony (Unfinished).* He played that record endlessly.

"Curtis," I asked, "why do you listen to that stuff? It sounds absolutely ridiculous."

"I like it," he said. He might have tried to explain a little about the music, but at the time I wasn't quite ready to hear him. However, he played that record so often during his two weeks at home that I found myself going around humming the melody. About that time I realized that I had actually begun to enjoy classical music!

Classical music wasn't totally foreign to me. I had

taken clarinet lessons since the seventh grade because that's what my brother played. And after all, that meant my mother had to rent only one instrument in the beginning, and I could use Curtis's old music. Later I went on to cornet until, in ninth grade, I switched to the baritone.

Curtis helped me to enjoy Schubert, and then I bought a record as a gift for my mother. Truthfully, I bought it for myself. The record contained the many overtures from Rossini's operas, including the most well-known *The William Tell Overture.*

My next step was listening to the German and the Italian arias. I read books about operas and understood the stories. By then I was saying, "This is great music." I no longer pushed myself to learn about classical music because I wanted to be on *College Bowl.* I had gotten hooked.

By the time I got to college I could listen to just about any piece of music—from classical to pop—and I'd know who wrote it. I have a good ear for recognizing styles in music, and I cultivated that.

During college, every evening I used to listen to a program called *The Top One Hundred.* It played only classical music. I listened every night, and it wasn't long before I knew the top one hundred cold. Then I decided to branch out from just classical music, so I made it a point to listen and learn from a wider range of music.

I did everything I knew to get ready to try out for the *College Bowl.* Unfortunately, I never did get to appear on the program.

---

[1] Curtis graduated from high school at the height of the war in Viet Nam. In those days the Selective Service used a lottery system to determine who should go into the military service. Curtis's low lottery number assured him that if he waited, the Army would draft him. After completing a year and a half of college, he decided to join the Navy. "I may as well get the branch of service that I want," he said.

He got into a special program, and the Navy trained him to be a nuclear submarine operator. It was a six-year program (although he did not re-enlist after his four-year stint). He progressed quite well through the ranks and probably would have been at least a captain by now if he had stayed in. However, he decided to go back to college. Today Curtis is an engineer, and I'm still proud of my big brother.

[2] I made second lieutenant after only three semesters when it usually took at least four, and most ROTC cadets never reached that rank in six semesters.

8 COLLEGE CHOICES

I stared at the ten-dollar bill on the table before me, knowing I had to make a choice. And since I had only one chance, I wanted to make sure I made the right one.

For days I'd considered the matter from every possible angle. I'd prayed for God to help me. But it still seemed to come down to making one single decision.

An ironic situation faced me in the fall of 1968, for most of the top colleges in the country had contacted me with offers and enducements. However, each college required a ten-dollar non-returnable entrance fee sent with the application. I had exactly ten dollars, so I could apply only to one.

Looking back I realize that I could have borrowed the money to make several applications. Or, it's possible that if I'd talked to representatives from the schools they might have waived the fee. But my mother had pushed the concept of self-reliance for so long I didn't want to start out owing a school just to get accepted.

At that time the University of Michigan—a spectacular school and always in the top ten academically and in sports events—actively recruited Black students. And the University of Michigan waived the fees for in-state students who couldn't afford to pay. However, I wanted to attend college farther away.

I looked hard at my future, knowing that I could get into any of the top schools but not knowing what to do. Graduating third in my class, I had excellent SAT scores, and most of the top colleges were scrambling to enroll Blacks. After college, with a major in premed and a minor in psychology, I'd be ready for medical school, and at last on the real road toward becoming a doctor.

For a long time it bothered me that I had graduated third in my senior high school class. It's probably a character flaw, but I can't help myself. It wasn't that I had to be first in everything, but I *should* have been number one. If I hadn't gotten so sidetracked by the need for peer approval, I would have been at the head of my class. In thinking toward college, I determined that would never happen again. From now on, I'd be the best student I was capable of being.

Several weeks flew by as I struggled over which college to send my application to, and by late spring I had narrowed the choice between Harvard and Yale. Either would have been great, which made the decision difficult. Strangely enough, my final decision hinged on a television program. As I watched *College Bowl* one Sunday night, the Yale students wiped the Harvard students off the face of the map with a fantastic score of something like 510 to 35. That game helped me to make my decision—I wanted to go to Yale.

In less than a month I not only had my acceptance at Yale to enter in the fall of 1969, but they offered me a 90 percent academic scholarship.

I suppose I should have been elated by the news. I was happy, but not surprised. Actually I took it calmly, and perhaps even a bit arrogantly, reminding myself that I had already accomplished just about everything I'd set out to do—a high scholastic record, top SAT scores, every kind of high school recognition possible, along with my long list of achievements with the ROTC program.

Campus accommodations befitted students of my stature. The student housing was luxurious, the rooms more like suites. The suites included a living room, fireplace, and built-in bookcases. Bedrooms branched off from the main room. Two to four students shared each suite. I had a room to myself.

I strode onto the campus, looked up at the tall, gothic-style buildings, and approved of the ivy-covered walls. I figured I'd take the place by storm. And why not? I was incredibly bright.

After less than a week on campus I discovered I wasn't that bright. All the students were bright; many of them extremely gifted and perceptive. Yale was a great leveler for me, because I now studied, worked, and lived with dozens of high-achieving students, and I didn't stand out among them.

One day I was sitting at the dining room table with several class members who were talking about their SAT scores. One of them said, "I blew the SAT test with a total of just a little over fifteen hundred in both parts."

"That's not too bad," another one sympathized. "Not great, but not bad."

"What did you get?" the first student asked him.

"Oh, 1540 or 1550, total. I can't remember my exact math score."

It seemed perfectly natural to all of them to have scores in the high ninety percentile. I kept silent, realizing that I ranked lower than every student sitting around me. It was my first awareness of not being quite as bright as I thought, and the experience washed away a little of my cockiness. At the same time, the incident only slightly deterred me. It would be simple enough to show them. I'd do what I did at Southwestern and throw myself completely into my studies, learning as much as possible. Then my grades would put me right up in the top echelon.

But I quickly learned that the classwork at Yale was difficult, unlike anything I'd ever encountered at Southwestern High School. The professors expected us to have done our homework before we came to class, then used that information as the basis for the day's lectures. This was a foreign concept to me. I'd slid through semester after semester in high school, studying only what I wanted, and then, being a good crammer, spent the last few days before exams memorizing like mad. It had worked at Southwestern. It was a shock to realize it wouldn't work at Yale.

Each day I slipped farther and farther behind in my classwork, especially in chemistry. Why I didn't work to keep up, I'm not sure. I could give myself a dozen excuses, but they didn't matter. What mattered was that I didn't know what was going on in chemistry class.

It all came to a head at the end of the first semester when I faced final examinations. The day before the exam I wandered around the campus, sick with dread. I couldn't deny it any longer. I was failing freshman chemistry; and failing it badly. My feet scuffed through the golden leaves carpeting the wide sidewalks. Sunlight and shadow danced on ivy-covered walls. But the beauty of that autumn day mocked me. I'd blown it. I didn't have the slightest hope of passing chemistry, because I hadn't kept up with the material. As the realization sunk in of my impending failure, this bright boy from Detroit also stared squarely into another horrible truth—if I failed chemistry I couldn't stay in the premed program.

Despair washed over me as memories of fifth grade flashed through my mind. "What score did you get, Carson?" "Hey, dummy, did you get any right today?" Years had passed, but I could still hear the taunting voices in my head.

What am I doing at Yale anyway? It was a legitimate question, and I couldn't push the thought away. *Who do I think I am? Just a dumb Black kid from the poor side of Detroit who has no business trying to make it through Yale with all these intelligent, affluent students.* I kicked a stone and sent it flying into the brown grass. *Stop it,* I told myself. *You'll only make it worse.* I turned my memories back to those teachers who told me, "Benjamin, you're bright. You can go places."

There, walking alone in the darkness of my thoughts, I could hear Mother insist, "Bennie, you can do it! Why, son, you can do anything you want, and you can do it better than anybody else. I believe in you."

I turned and began walking between the tall, classic buildings back to the dorm. I had to study. *Stop thinking about failing,* I told myself. *You can still pull this off. Maybe.* I looked up through a scatter of fluttering leaves silhouetted against the rosy autumn sunset. Doubts niggled at the back of my mind.

Finally I turned to God. "I need help," I prayed. "Being a doctor is all I've ever wanted to do, and now it looks like I can't. And, Lord, I've always had the impression You wanted me to be a doctor. I've worked hard and focused my life that way, assuming that's what I was going to do. But if I fail chemistry I'm going to have to find something else to do. Please help me know what else I should do."

Back in my room, I sank down on my bed. Dusk came early, and the room was dark. The evening sounds of campus filled the quiet room—cars passing, students' voices in the park below my window, gusts of wind rustling through the trees. Quiet sounds. I sat there, a tall, skinny kid, head in my hands. I had failed. I had finally faced a challenge I couldn't overcome; I was just too late.

Standing up, I flipped on the desk lamp. "OK," I said to myself as I paced my room, "I'm going to fail chemistry. So I'm not going to be a doctor. Then what is there for me?"

No matter how many other career choices I considered, I couldn't think of anything else in the whole world I wanted more than being a doctor. I remembered the scholarship offer from West Point. A teaching career? Business? None of these areas held any real interest.

My mind reached toward God—a desperate yearning, begging, clinging to Him. "Either help me understand what kind of work I ought to do, or else perform some kind of miracle and help me to pass this exam."

From that moment on, I felt at peace. I had no answer. God didn't break through my haze of depression and flash a picture in front of me. Yet I knew that whatever happened, everything was going to be all right.

One glimmer of hope—a tiny one at that—shone through my seemingly impossible situation. Although I had been holding on to the bottom rung of the class from the first week at Yale, the professor had a rule that might save me. If failing students did well on the final exam, the teacher would throw out most of the semester's work and let the good final-test score count heavily toward the final grade. That presented the only possibility for me to pass chemistry.

It was nearly 10:00 p.m., and I was tired. I shook my head, knowing that between now and tomorrow morning I couldn't pull off that kind of miracle.

"Ben, you have to try," I said aloud. "You have to do everything you can."

I sat down for the next two hours and pored through my thick chemistry textbook, memorizing formulas and equations that I thought might help. No matter what happened during the exam, I would go into it determined to do the best I could. I'd fail but, I consoled myself, at least I'd have a high fail.

As I scribbled formulas on paper, forcing myself to memorize what had no meaning to me, I knew deep inside why I was failing. The course wasn't that tough. The truth lay in something much more basic. Despite my impressive academic record in high school, I really hadn't learned anything about studying. All the way through high school I'd relied on the same old methods —wasting my time during the semester, and then cramming for final exams.

Midnight. The words on the pages blurred, and my mind refused to take in any more information. I flopped into my bed and whispered in the darkness, "God, I'm sorry. Please forgive me for failing You and for failing myself." Then I slept.

While I slept I had a strange dream, and, when I awakened in the morning, it remained as vivid as if it had actually happened. In the dream I was sitting in the chemistry lecture hall, the only person there. The door opened, and a nebulous figure walked into the room, stopped at the board, and started working out chemistry problems. I took notes of everything he wrote.

When I awakened, I recalled most of the problems, and I hurriedly wrote them down before they faded from memory. A few of the answers actually did fade but, still remembering the problems, I looked them up in my textbook. I knew quite a bit about psychology so assumed I was still trying to work out unresolved problems during my sleep.

I dressed, ate breakfast, and went to the chemistry lecture room with a feeling of resignation. I wasn't sure if I knew enough to pass, but I was numb from intensive cramming and despair. The lecture hall was huge, filled with individual fold-down wooden seats. It would seat about 1,000 students. In the front of the room chalkboards faced us from a large stage. Also on the stage was a big desk with a countertop and sink for chemistry demonstrations. My steps sounded hollow on the wooden floor.

The professor came in and, without saying much, began to hand out the booklets of examination questions. My eyes followed him around the room. It took him a while to pass out the booklets to 600 students. While I waited, I noticed the way the sun shone through the small panes of the arched windows along one wall. It was a beautiful morning to fail a test.

At last, heart pounding, I opened the booklet and read the first problem. In that instant, I could almost hear the discordant melody that played on TV with *The Twilight Zone*. In fact, I felt I had entered that never-never land. Hurriedly I skimmed through the booklet, laughing silently, confirming what I suddenly knew. The exam problems were identical to those written by the shadowy dream figure in my sleep.

I knew the answer to every question on the first page. "Piece of cake," I mumbled as my pencil flew to write the solutions. The first page finished, I turned to the next page, and again the first problem was one I had seen written on the board in my dream. I could hardly believe it.

I didn't stop to analyze what was happening. I was so excited to know correct answers that I worked quickly, almost afraid I'd lose what I remembered. Near the end of the test, where my dream recall began to weaken, I didn't get every single problem. But it was enough. I knew I would pass.

"God, You pulled off a miracle," I told Him as I left the classroom. "And I make a promise to You that I'll never put You into that situation again."

I walked around campus for over an hour, elated, yet needing to be alone, wanting to figure out what had happened. I'd never had a dream like that before. Neither had anyone I'd ever known. And that experience contradicted everything I'd read about dreams in my psychological studies.

The only explanation just blew me away. The one answer was humbling in its simplicity. For whatever reason, the God of the universe, the God who holds galaxies in His hands, had seen a reason to reach down to a campus room on Planet Earth and send a dream to a discouraged ghetto kid who wanted to become a doctor.

I gasped at the sure knowledge of what had happened. I felt small and humble. Finally I laughed out loud, remembering that the Bible records such events, though they were few—times where God gave specific answers and directions to His people. God had done it for me in the twentieth century. Despite my failure, God had forgiven me and come through to pull off something marvelous for me.

"It's clear that You want me to be a doctor," I said to God. "I'm going to do everything within my power to be one. I'm going to learn to study. I promise You that I'll never do this to You again."

During my four years at Yale I did backslide a little, but never to the point of not being prepared. I started learning how to study, no longer concentrating on surface material and just what the professors were likely to ask on finals. I aimed to grasp everything in detail. In chemistry, for instance, I didn't want to know just answers but to understand the reasoning behind the formulas. From there, I applied the same principle to all my classes.

After this experience, I had no doubt that I would be a physician. I also had the sense that God not only wanted me to be a physician, but that He had special things for me to do. I'm not sure people always understand when I say that, but I had an inner certainty that I was on the right path in my life—the path God had chosen for me. Great things were going to happen in my life, and I had to do my part by preparing myself and being ready.

When the final chemistry grades came out, Benjamin S. Carson scored 97—right up there with the top of the class.

9 CHANGING THE RULES

During my college years I worked at several different summer jobs, a practice I had started in high school where I worked in the school laboratory. The summer between my junior and senior year of high school, I worked at Wayne State University in one of the biology laboratories.

Between high school graduation and entering Yale I needed a job badly. I had to have clothes for college, books, transportation money, and the dozens of other expenses I knew I'd face.

One of the counselors at our high school, Alma Whittley, knew my predicament and was very understanding. One day I poured out my story, and she listened with obvious concern. "I've got a few connections with the Ford Motor Company," she said. While I sat next to her desk, she phoned their world headquarters. I particularly remember her saying, "Look, we have this young fellow here named Ben Carson. He's very bright and already has a scholarship to go to Yale in September. Right now the boy needs a job to save money for this fall." She paused to listen, and I heard her add, "You have to give him a job."*

The person on the other end agreed.

The day after my last high school class my name went on the list of employees at the Ford Motor Company in the main administration building in Dearborn. I worked in the payroll office, a job I considered prestigious, or as my mother called it, big time, because they required me to wear a white shirt and tie every day.

That job taught me an important lesson about employment in the world beyond high school. Influence could get me inside the door, but my productivity and the quality of my work were the real tests. Just knowing a lot of information, while helpful, wasn't enough either. The principle goes like this: It's not what you know but the kind of job you do that makes the difference.

That summer I worked hard, as I did at every job, even the temporary ones. I determined that I would be the best person they had ever hired.

After completing my first year at Yale, I received a wonderful summer job as a supervisor with a highway crew—the people who clean up the trash along the highways. The federal government had set up a jobs program, mostly for inner-city students. The crew walked along the Interstate near Detroit and the western suburbs, picking up and bagging trash in an effort to keep the highways beautiful.

Most of the supervisors had a horrible time with discipline problems, and the inner-city kids had hundreds of reasons for not putting any effort into their work. "It's too hot to work today," one would say. "I'm just too tired out from yesterday," another said. "Why we gotta do all this? Tomorrow people will just litter it all up again. Who'll know if we cleaned it up or not?" "Why should we kill ourselves at this? The job just doesn't pay enough to do that."

The other supervisors, I learned, figured that if each of the five to six young men in the crew filled two plastic bags a day, they were doing well.

These guys could do that much in one hour, and I knew it. I may be an overachiever, but it seemed a waste of my time to let my crew laze around picking up 12 bags of litter a day. From the first my crew consistently filled between 100 and 200 bags a day, and we covered enormous stretches of highway.

The amount of work my crew did flabbergasted my supervisiors in the Department of Public Works. "How come your guys can get so much work done?" they asked. "None of the other crews do that much."

"Oh, I have my little secrets," I'd say, and make a joke out of what I was doing. If I said too much, someone might interfere and make me change my rules.

I used a simple method, but I didn't go by the standard procedures—and I share this story because I think it illustrates another principle in my life. It's like the popular song of a few years ago that says "I did it my way." Not because I oppose rules—it would be crazy to to do surgery without obeying certain rules—but some-times regulations hinder and need to be broken or ignored.

For example, the fourth day on the job I said to my guys, "It's going to be real hot today—"

"You can say that again!" one of them said, and immediately they all eagerly agreed.

"So," I said, "I'm going to make you a deal. First, beginning tomorrow, we start at six in the morning while it's still cool—"

"Man, nobody in the whole world gets up that early—"

"Just listen to my whole plan," I said to the inter-rupter. Our crews were supposed to work from 7:30 a.m. until 4:30 p.m. with an hour off for lunch. "If you guys—and it has to be all six of you—will be ready to start work so that we can get out on the road at six, and you work fast to fill up 150 bags, then after that you're through for the day." Before anyone could start ques-tioning me I clarified what I meant.

"You see, if you can collect all that trash in two hours, I'll take you back, and you're off the rest of the day. You still earn a full day's pay. But you have to bring in 150 bags no matter how long it takes."

We bashed the idea back and forth, but they saw what I wanted. It had only taken a couple of days to get them to pick up 100 bags a day, and it was hot, hard work in the afternoon. But they loved taunting the other crews and telling how much they had done, and they were ready for the new challenge. These kids were learning to take pride in their work, as lowly as many of them considered their jobs.

They agreed with my arrangement. The next morning all six of them were ready to go at 6:00 a.m. And how they worked—hard and fast. They learned to clean a whole stretch of highway in two to three hours—the same amount of work that they had previously stretched out for the whole day.

"OK, guys," I'd say as soon as I counted the last bag. "We take the rest of the day off."

They loved it and worked with a joyful playfulness. Their best moments came when we'd be hauling ourselves into the Department of Transportation by 9:00, just as the other crews were getting started.

"You guys going to work today?" one of my guys would yell.

"Man, not much trash out there today," another one would say. "Superman and his hot shots have cleaned up most of it."

"Hope you don't get sunburned out there!" they yelled as a truck pulled out.

Obviously the supervisors knew what I was doing, because they saw us coming back in, and they certainly had reports of our going out early. They never said anything. If they had, all I would have had to do was produce evidence of our work.

We weren't supposed to work that way, because the rules set the specific work hours. Yet not one supervisor ever commented on what I was doing with my crew. More than anything else, I believe they kept silent because we were getting the job done and doing it faster and better than any of the other crews.

Some people are born to work, and others are pushed into it by their moms. But doing what must be done as quickly and as well as possible has been my strategy for everything, including medicine. We don't necessarily have to play by the strict rules if we can find a way that works better, as long as it's reasonable and doesn't hurt anybody. Someone told me that creativity is just learning to do something with a different perspective. So maybe that's what it is—being creative.

The following summer, after my second year of college, I came back to Detroit to work again as a supervisor with my road crew. At the end of the previous year, Carl Seufert, the top man in the Department of Transportation, had left me with the words "Come on back next summer. We'll have a place for you."

However, the economy hit a slump in the summer of '71, especially in the capital of the automobile industry. Supervisory positions, because they paid well, were incredibly hard to get. Most of the college students who got those jobs had significant personal or political connections. They had been hired months in advance while I was still in New Haven.

Since Carl Seufert had promised me a job, I didn't consider confirming it during the Christmas vacation period. When I applied in late May, the personnel director said, "I'm sorry. Those jobs are all gone." She explained the situation of few jobs and more applicants, but I already knew that.

I didn't blame that woman, and I knew arguing with her wouldn't get me anywhere. I should have put in my application earlier like the others.

But I confidently reasoned that I had worked every summer, and I would find another job easily enough.

I was wrong. Like hundreds of college students, I found that there were absolutely no jobs anywhere. I beat the streets for two weeks. Each morning I'd get on the bus, ride downtown, and apply at every business establishment I came across.

"Sorry, no jobs." I must have heard that statement, or variations of it, a hundred times. Sometimes I heard genuine sympathy in the voice that said it. At other places, I felt as if I was number 8,000 to come in, and the person was tired of repeating the same thing and just wished we'd all go away.

In the middle of this depressing search for employment, Ward Randall, Jr., was a bright light in my life.

Ward, a White attorney in the Detroit area, had graduated from Yale two decades before me. We met at a local alumni meeting while I was still a student. He took a liking to me because we both shared a keen interest in classical music. During the summer of 1971 when I was searching for a job in downtown Detroit, we frequently met for lunch and then went to the noonday concerts. Many of them were organ concerts in one of the churches downtown.

Besides that, Ward frequently invited me to go with his family to various concerts and symphonies, and he introduced me to a lot of the cultural interests around Detroit that I wouldn't have had the opportunity to attend because of my lack of finances. He was just a real nice man, a real encouragement to me, and I still appreciate him today.

After walking all over the city, I finally decided, *I'm going to make up my own rules on this one. I've tried all the conventional ways of finding a job and found nothing. Nothing. Nothing.*

Then I remembered my regional interview for entrance into Yale and the person who had interviewed me—a nice man named Mr. Standart. He was also the vice president of Young and Rubicum Advertising, one of the large national advertising companies.

First I tried the personnel office of his company and received the familiar words "I'm sorry, we have no temporary jobs available."

Casting aside my pride and giving myself another pep talk, I got on the elevator to the executive suites. Because Mr. Standart had interviewed me for Yale and given me a fine recommendation, I figured he must have had a good opinion of me. But I hadn't figured out how I'd get past his secretary. I remembered that nobody, absolutely nobody, got into his office without an appointment. Then I figured, "What have I got to lose?"

When Mr. Standart's secretary looked up at me, I said, "My name is Ben Carson. I'm a student from Yale, and I'd like to see Mr. Standart for just a minute—"

"I'll see if he's free." She went into his office, and a minute later Mr. Standart himself came out. He smiled, and his eyes met mine as he held out his hand. "Nice of you to come by and see me," he said. "How are things going for you at Yale?"

As soon as we finished the formalities, I said, "Mr. Standart, I need a job. I'm having a terrible time trying to find work. I've been out every day for two weeks, and I can't find a thing."

"Is that right? Did you try personnel here?"

"No jobs here either," I said.

"We'll just have to see what we can do." Mr. Standart picked up the phone and punched a couple of numbers, while I looked around his mammoth office. It was exactly like the fabulous sets of executive suites I'd seen on television.

I didn't hear the name of the person he talked to, but I heard the rest of his words. "I'm sending a young man down to your office. His name is Ben Carson. Find a job for him."

Just that. Not given as a harsh command but as a simple directive from the kind of man who had the authority to issue that kind of order.

After thanking Mr. Standart I went back to the personnel office. This time the director of personnel himself talked to me. "We don't need anybody, but we can put you in the mail room."

"Anything. I just need a job for the rest of the summer."

The job turned out to be a lot of fun because I got to drive all around the city, delivering and picking up letters and packages.

I had only one problem. The job just didn't pay enough for me to save anything for school. After three weeks, I took my next step of action. I decided that I had to quit my job and find one that paid better. "After all,"

I said to reinforce my decision, "it worked with Mr. Standart." I went to the Department of Transportation and talked to Carl Seufert.

We were already nearing the end of June, every job was filled, and it seemed pretty audacious for me to try, but I did it anyway.

I went directly to Mr. Seufert's office, and he had time to talk to me. After he heard my summer's tale, he said, "Ben, for a guy like you there's always a job." He was the overall supervisor of the highway construction crews, both cleanup and highway maintenance. "Since the supervisory jobs are all gone," he said, "we'll make a job." He paused and thought for a few seconds and said, "We'll just set up another crew and give you a job."

That's exactly what Mr. Seufert did. By using creativity and a little daring, I got my old job back. I used the same tactics with my new six-member crew, and it worked as effectively as it had the previous summer.

Frequently I'd see Carl Seufert when I checked out, or he'd visit us on the worksite. He'd always take time to chat with me. "Ben," he said to me more than once, "you're a good man. We're fortunate to have you."

On one occasion he put his arm on my shoulder and said, "You're your own man. You can accomplish anything that you want in the world." As I listened, this man began to sound like my mother, and I loved hearing his words. "Ben, you're a talented person, and you can do anything. I believe you're going to do great things. I'm just glad to know you."

I've always remembered his words.

The following summer, 1972, I worked on the line for Chrysler Motor Company, assembling fender parts. Each day I went to work and concentrated on doing my best. Some may find this hard to believe, but with only three months on the job, I received recognition and promotion. Toward the end of the summer they moved me up to inspect the louvers that go on the back windows of the sporty models. I got to drive some of the cars off the finish line to the place where we parked them for transportation to showrooms. I liked the things

I did at Chrysler. And every day there confirmed what I had already believed.

That summer I also learned a valuable lesson—one that I'd never forget. My mother had given me the words of wisdom, but, like many kids, I paid little attention. Now I knew from my own experience how right she was: The kind of job doesn't matter. The length of time on the job doesn't matter, for it's true even with a summer job. If you work hard and do your best, you'll be recognized and move onward.

Although said a little differently, my mother had given me the same advice. "Bennie, it doesn't really matter what color you are. If you're good, you'll be recognized. Because people, even if they're prejudiced, are going to want the best. You just have to make being the best your goal in life."

I knew she had been right.

———

Lack of money constantly troubled me during my college years. But two experiences during my studies at Yale reminded me that God cared and would always provide for my needs.

First, during my sophomore year I had very little money. And then all of a sudden, I had absolutely no money—not even enough to ride the bus back and forth to church. No matter how I viewed the situation, I had no prospects of anything coming in for at least a couple of weeks.

That day I walked across the campus alone, bewailing my situation, tired of never having enough money to buy the everyday things I needed; the simple things like toothpaste or stamps. "Lord," I prayed, "please help me. At least give me bus fare to go to church."

Although I'd been walking aimlessly, I looked up and realized I was just outside Battell Chapel on the old campus. As I approached the bike racks, I looked down. A ten-dollar bill lay crumpled on the ground three feet in front of me.

"Thank You, God," I said as I picked it up, hardly able to believe that I had the money in my hand.

The following year I hit that same low point again —not one cent on me, and no expectations for getting any. Naturally I walked across campus all the way to the chapel, searching for a ten-dollar bill. I found none.

Lack of funds wasn't my only worry that day, however. The day before I'd been informed that the final examination papers in a psychology class, Perceptions 301, "were inadvertently burned." I'd taken the exam two days earlier but, with the other students, would have to repeat the test.

And so I, with about 150 other students, went to the designated auditorium for the repeat exam.

As soon as we received the tests, the professor walked out of the classroom. Before I had a chance to read the first question, I heard a loud groan behind me.

"Are they kidding?" someone whispered loudly.

As I stared at the questions, I couldn't believe them either. They were incredibly difficult, if not impossible. Each of them contained a thread of what we should have known from the course, but they were so intricate that I figured a brilliant psychiatrist might have trouble with some of them.

"Forget it," I heard one girl say to another. "Let's go back and study this. We can say we didn't read the notice. Then when they repeat it, we'll be ready." Her friend agreed, and they quietly slipped out of the auditorium.

Immediately three others packed away their papers. Others filtered out. Within ten minutes after the exam started, we were down to roughly one hundred. Soon half the class was gone, and the exodus continued. Not one person turned in the examination before leaving.

I kept working away, thinking all the time, *How can they expect us to know this stuff?* Pausing then to look around, I counted seven students besides me still going over the test.

Within half an hour from the time the examination began, I was the only student left in the room. Like the others, I was tempted to walk out, but I had read the notice, and I couldn't lie and say I hadn't. All the time I wrote my answers, I prayed for God to help me figure out what to put down. I paid no more attention to departing footsteps.

Suddenly the door of the classroom opened noisily, disrupting my flow of thought. As I turned, my gaze met that of the professor. At the same time I realized no one else was still struggling over the questions. The professor came toward me. With her was a photographer for the Yale *Daily News* who paused and snapped my picture.

"What's going on?" I asked.

"A hoax," the teacher said. "We wanted to see who was the most honest student in the class." She smiled again. "And that's you."

The professor then did something even better. She handed me a ten-dollar bill.

---

* In the summer of 1988 Mrs. Whittley sent me a note that started out, "I wonder if you remember me." I was touched and tickled. Of course I remembered her, as I would have remembered anyone who had been that helpful to me. She said she had seen me on television and read articles about me. She is now retired, living in the South, and she wanted to send me her congratulations.

I was delighted that *she* remembered me.

10 A SERIOUS STEP

I've always been called Candy," she said, "but my name is Lacena Rustin."

Momentarily I stared, mesmerized by her smile. "Nice to meet you," I replied.

She was one of many freshmen I met that day at the Grosse Pointe Country Club. Many of Michigan's wealthiest citizens live in Grosse Pointe, and tourists often come to admire the homes of the Fords and Chryslers. Yale was hosting a freshmen reception for new students, and I, along with a number of upperclassmen, attended to welcome students from Michigan. It had meant a lot to me to have some connections when I first went away to college, and I enjoyed meeting and helping the new students whenever I could.

Candy was pretty. I remember thinking *That's one good looking girl.* She had an exuberance about her that I liked. She was bubbly, sort of all over the place, talking to this one and that. She laughed easily, and during the few minutes that we talked she made me feel good.

At five feet seven, Candy was about half a foot

shorter than I am. Her hair fluffed around her face in the popular Afro style. But most of all, I was drawn to her effervescent personality. Maybe because I tend to be quiet and introspective, and she was so outgoing and friendly, I admired her from the start.

At Yale, mutual friends often said, "Ben, you ought to get together with Candy." I later found out that friends would say to her, "Candy, you and Ben Carson ought to get together. You just seem right together."

Though I was beginning my third year of college when we met, I definitely wasn't ready for love. With my lack of finances, my single-minded goal to become a doctor, and the long years of study and internship that I faced, falling in love was the last thing on my mind. I'd come too far to get sidetracked by romance. Another factor entered into the picture, too. I'm rather shy and hadn't done much dating. I'd gone out with small groups, dated now and then, but had never gotten into any serious relationship. And I didn't plan on any either.

Once school began, I saw Candy occasionally since we were both in the premed program. "Hi," I'd call out. "How are you doing in your classes?"

"Fantastic," she'd usually say.

"You're adjusting all right then?" I asked the first time.

"I think I'm going to get straight A's."

As we chatted I'd think, *This girl must be really smart.* And she was.

I was even more amazed when I learned that she played violin in the Yale Symphony and Bach Society—not a position for just anybody who could play an instrument. These folks were top-grade musicians. As the weeks and months passed by, I learned more and more intriguing things about Candy Rustin. The fact that she was musically talented and knew classical music gave us something to talk about as we'd pass from time to time on campus.

However, Candy was just another student, a nice person, and I didn't have any particularly warm feelings toward her. Or perhaps, with my head in my books and my sights set on medical school, I wouldn't let myself consider how I really felt about the bright and talented Candy Rustin.

About the time Candy and I started talking more often and for longer periods, the church in New Haven which I attended needed an organist.

I had mentioned our choir director, Aubrey Tompkins, to Candy several times, because he was an important part of my life. After I joined the church choir, Aubrey would come by and pick me up on Friday evenings for rehearsal. During my second year my roommate Larry Harris, who was also an Adventist, joined the choir. Often on Saturday nights Aubrey took Larry and me to his home, and we grew to know his family well. At other times he showed us the sights of New Haven. An opera buff, Aubrey invited me several times to go with him on Saturday nights to the Metropolitan Opera in New York.

"Say, Candy," I told her one day, "I just thought of something. You're a musician. Our church needs an organist. What do you think? Would you be interested in the job? They pay the organist, but I don't know how much."

She didn't even hesitate. "Sure," she said, "I'd like to try it."

Then I paused with a sudden thought. "Do you think you could play the music? Aubrey gives us some difficult stuff."

"I can probably play anything with practice."

So I told Aubrey Tompkins about Candy. "Fantastic!" he responded. "Have her come for an audition."

Candy came to the next choir rehearsal and played the large electric organ. She played well, and I was happy just to see her up there, but the violin was her instrument. She could play anything written for the violin. And although Candy had played the organ for her high school baccalaureate service, she hadn't had much of an opportunity to keep in practice. She had no idea that Aubrey Tompkins liked to throw us into the heavy stuff, particularly Mozart, and she wasn't quite up to it on the organ.

Aubrey let her play a few minutes; then he said kindly, "Look, dear, why don't you sing in the choir?"

She could have had her feelings hurt, but Candy had enough self-confidence to take it in stride. A master on the violin, the organ wasn't her principal instrument. "All right," she said. "I guess I'm not so hot on the organ."

So Candy walked over to where we were singing and joined in. She had a lovely alto voice. And I was delighted when she joined us. She was a real addition to the choir. Everyone loved her from that first night, and, because she liked singing with us, Mt. Zion became Candy's church too from then on.

She wasn't overly religious, didn't talk much about spiritual or religious things, and had no significant Biblical background. But she was open and ready to learn.

After Candy started attending our church, she enrolled in special Bible classes that lasted from autumn to spring. I used to go with her one or two nights each week, learning a great deal about the Bible myself, and enjoying her company at the same time.

As Candy reflects on her spiritual journey, she says she always seemed to have a hunger for God. But what made it different for her in the Adventist church? "The people," she says. "They loved me into the faith."

Her family thought it was strange for her to join with Christians who went to church on Saturday. Yet eventually they not only accepted her decision, but Candy's mother became an active Adventist herself.

———

Candy and I soon fell into the habit of meeting each other after class. We walked across campus together or occasionally went into New Haven.

I was beginning to like Candy a lot.

Just before Thanksgiving of 1972, when I was in my final year at Yale and Candy was a sophomore, the admissions office paid our way to do recruiting in the high schools in the Detroit area. They provided us with an expense account, so I rented a little Pinto, and with our extra money we were able to eat in several nice restaurants. It was just the two of us, and we had a wonderful time.

We spent a lot of time together and the reality slowly came to me that I liked Candy quite a lot. More than I'd been aware of; more than I'd ever liked a girl.

Yale had recruited Candy and me to interview students who had combined SATs of at least twelve hundred. After going to all the inner-city schools in Detroit, we didn't find one student who had a combined SAT score to reach that total. To interview any students, Candy and I had to visit places in the more affluent communities like Bloomfield Hills and Grosse Pointe. We found plenty of students to interview who wanted to talk about attending Yale, but we didn't recruit any minorities.

On the trip Candy met my mother and some of my friends. Consequently, we ended up staying a little longer in Detroit than I had planned. I needed to have the rented Pinto back at the agency by 8:00 the next morning. That meant we had to drive straight through from Detroit.

The weather had been cold. A light snow had fallen the day before, although most of it had melted. Since leaving Yale ten days earlier, I hadn't once had an adequate night's sleep, because of our work and wanting to spend time with friends.

"I don't know if I can stay awake," I told Candy with a yawn. Most of the driving would be on the interstate highways, which makes driving monotonous.

Candy and I later disagreed on how she answered. I thought she said something like, "Don't worry, Ben, I'll keep you awake." She hadn't had any more sleep than I had. She says her words were, "Don't worry, Ben, you'll stay awake."

We started back to Connecticut. Back then, the speed limit was 70 miles per hour, but I must have been hitting close to 90. And what could be more boring to my sleep-starved body than watching endless median marks flashing by on a dark, moonless night?

By the time I crossed the line into Ohio, Candy had drifted off to sleep, and I didn't have the heart to awaken her. Though we'd had a wonderful time, the days away from school had been hard on both of us, and I figured that maybe she'd rest a couple of hours, then be fully awake and take over the wheel.

About one in the morning I was zooming along Interstate 80 and recall passing a sign that indicated we were nearing Youngstown, Ohio. With my hands relaxed on the wheel, the car flew along at 90 miles per hour. The heater, turned on low, kept us comfortably warm. It had been half an hour or more since I'd seen another vehicle. I felt relaxed, everything under control.

Then I floated into a comfortable sleep too.

The vibration of the car striking the metal illuminators that separate each lane jarred me into consciousness. My eyes popped opened as the front tires struck the gravel shoulder. The Pinto veered off the road, the headlights streaming into the blackness of a deep ravine. I yanked my foot off the gas pedal, grabbed the steering wheel, and fiercely jerked to the left.

In those action-packed seconds, my life flashed before my eyes. I'd heard people say that a slow-motion review of life tumbles through the mind just before one dies. *This is a prelude to death,* I thought. *I'm going to die.* A panorama of experiences from early childhood to the present rolled across my mind. *This is it. This is the end.* The words kept rumbling through my head.

Going at that speed, the car should have flipped over, but a strange thing happened. Because of my overcorrection with the steering wheel, the car went into a crazy spin, around and around like a top. I released the wheel, my mind fully concentrating on being ready to die.

Abruptly the Pinto stopped—in the middle of the lane next to the shoulder—headed in the right direction, the engine still running. Hardly aware of what I was doing, my shaking hands slowly turned the wheel and pulled the car off onto the shoulder. A heartbeat later an eighteen-wheeler transport came barreling through on that lane.

I cut off the ignition and sat quietly, trying to breathe normally again. My heart felt as if it were racing at 200 beats a minute. "I'm alive!" I kept repeating. "Praise the Lord. I can't believe it, but I'm alive. Thank You, God. I know You've saved our lives."

Candy must have really been tired, for she'd slept through the whole terrible experience. My voice reached inside her sleep, though, and she opened her eyes. "Why are we parked here? Anything wrong with the car?"

"Nothing's wrong," I said. "Go back to sleep."

There must have been an edge to my voice, for she said, "Don't be like that, Ben. I'm sorry I fell asleep—I didn't mean to—"

I took a deep breath. "Everything's fine," I said and smiled at her through the darkness.

"Everything can't be fine if we're not moving. What's going on? Why are we stopped?"

I leaned forward and flipped on the ignition. "Oh, just a quick rest," I said casually, as I began to accelerate and pull onto the road.

"Ben, please—"

With a mixture of fear and relief, I let the car come to a stop far onto the road shoulder and turned off the key. "OK," I sighed. "I fell asleep back there . . ." My heart still pounded, my muscles were tense as I told her what happened. "I thought we were going to die," I concluded. I could hardly say the last words aloud.

Candy reached across the seat and put her hand in mine. "The Lord spared our lives. He's got plans for us."

"I know," I said, feeling just as certain of that fact as she did.

Neither of us slept the rest of the trip. We talked the whole time, the words flowing naturally between us.

At one point Candy said, "Ben, why are you always so nice to me? Like tonight. I did go to sleep when I probably should have stayed awake and talked to you."

"Well, I'm just a nice guy."

"It's more than that, Ben."

"Oh, I like being nice to second-year Yale students."

"Ben. Be serious."

The first brush of violet painted the horizon. I looked straight ahead, both hands on the wheel. Something unfamiliar fluttered in my chest as Candy persisted.

"Why?" It was hard to stop joking, hard to let the mask fall away and say the actual words. "I guess," I said, "it's because I like you. I guess I like you a lot."

"I like you a lot too, Ben. More than anybody else I've ever met."

I didn't answer but let the car slow down, eased it off the road, and stopped. It took only a moment to put my arms around Candy and kiss her. It was our first kiss. Somehow I knew she'd kiss me back.

We were two naive kids, and neither of us knew much about dating or carrying on a romance. But we both understood one thing—we loved each other.

From then on, Candy and I were inseparable, spending every possible minute together. Oddly enough, our growing relationship didn't detract me from my studies. Having Candy at my side, always encouraging me, made me just that more determined to work hard.

Candy didn't shirk her studies either. She was a triple major, carrying enough courses for music, psychology, and premed. Subsequently she dropped the premed to concentrate more on her music. Candy is one of the brightest people I know, good at whatever she does.*

———

One problem that bothered many in the premed program was getting into medical school after graduation. The system for medical training requires students to spend four years earning an undergraduate degree and then, if accepted by a medical school, to undergo another four years of intensive training.

"If I don't make it into med school," one of my classmates said several times, "I've just been wasting all this time."

"I don't know if I'll get in at Stanford," one premed said to me, after he had sent in his application. "Or anywhere else," he added.

Another mentioned a different school, but the students' worries were essentially the same. I seldom got involved in what I called freaking out, but this kind of talk happened often, especially during our senior year.

One time when this freaking out was going on and I didn't enter in, one of my friends turned to me. "Carson, aren't you worried?"

"No," I said. "I'm going to the University of Michigan Medical School."

"How can you be so sure?"

"It's real simple. My Father owns the university."

"Did you hear that?" he yelled at one of the others. "Carson's old man owns the University of Michigan."

Several students were impressed. And understandably, because they came from extremely wealthy homes. Their parents owned great industries. Actually I had been teasing, and maybe it wasn't playing fair. As a Christian I believe that God—my Heavenly Father—not only created the universe, but He controls it. And, by extension, God owns the University of Michigan and everything else.

I never did explain.

After graduating in 1973 from Yale, I ended up with a fairly respectable grade point average, although far from the top of the class. But, I knew I had done my best and put forth the maximum effort; I was satisfied.

Aside from my joking, I had no doubts about being accepted at the University of Michigan, Ann Arbor, in their School of Medicine. I applied there and since I had believed so strongly that God wanted me to be a doctor, I had no doubts about being accepted. Several of my friends wrote to half a dozen medical schools, hoping one would accept them. For two reasons I applied there and to only a few others. First, the University of Michigan was in my home state, which meant much lower school expenses for the next four years. Second, U of M had the reputation for being one of the top schools in the nation.

I had also applied to Johns Hopkins, Yale's medical school, Michigan State, and Wayne State. My acceptance from U of M came extremely early, so I immediately withdrew from the others. Candy still had two years of schooling at Yale when I began medical school, but we found ways to bridge time and space. We wrote to each other every single day. Even today both of us have boxes of love letters we saved.

When we could afford to, we used the telephone. One time I called her at Yale, and I don't know what happened, but neither of us seemed able to stop talking. Maybe we were both extra lonely. Maybe we'd both been having a hard time. Maybe we just needed to be together, to keep contact when our lives were so far apart. Anyway, we talked for six straight hours. At the time I didn't care. I loved Candy, and every second on the telephone was precious.

The next day I began to worry about paying the telephone bill. In one letter I joked about having to make payments all through my medical career. I wondered what the telephone company could do to a poor medical student who had even less sense than money.

I kept waiting and dreading the day when I actually saw the bill. Strangely enough, the 6-hour call never came through. I couldn't have paid it anyway—certainly not the whole amount—so I confess I didn't investigate the reason. As Candy and I talked it over later, we theorized that the phone company looked at the charges, and some executive decided that no one could possibly talk that long.

The summer between college graduation and medical school found me back to my old routine of hunting a job. And, as I had experienced before, I couldn't find any employment. This time I had started making contacts in the spring, three months before graduation. But Detroit was in the middle of an economic depression, and many employers said, "Hire you? Right now we're laying off people."

At that time my mother was caring for the children of the Sennet family—Mr. Sennet was the president of Sennet Steel. After hearing my sad tales, Mother told her employer about me. "He needs a job real bad," she said. "Is there any way you could help him?"

"Sure," he said. "I'd be happy to give your son a job. Send him over."

He hired me. I was the only one at Sennet Steel with a summer job. To my surprise, my foreman taught me how to operate the crane, a very responsible job, for it involved picking up stacks of steel weighing several tons. Whether he realized it or not, the operator had to have an understanding of physics to be able to visualize what he was doing as he moved the boom over and down to the steel. The immense stacks of steel had to be picked up in a certain way to prevent the bundles from swinging. Then the operator worked the crane to carry the steel over and into trucks that were parked in an extremely narrow space.

Somewhere during that period of time I became acutely aware of an unusual ability—a divine gift, I believe—of extraordinary eye and hand coordination. It's my belief that God gives us all gifts, special abilities that we have the privilege of developing to help us serve Him and humanity. And the gift of eye and hand coordination has been an invaluable asset in surgery. This gift goes beyond eye-hand coordination, encompassing the ability to understand physical relationships, to think in three dimensions. Good surgeons must understand the consequences of each action, for they're often not able to see what's happening on the other side of the area in which they're actually working.

Some people have the gift of physical coordination. These are the people who become Olympic stars. Others can sing beautifully. Some people have a natural ear for languages or a special aptitude for math. I know individuals who seem to draw friends, who have a unique ability to make people feel welcome and part of the family.

For some reason, I am able to "see" in three dimensions. In fact, it seems incredibly simple. It's just something I happen to be able to do. However, many doctors don't have this natural ability, and some, including surgeons, never learn this skill. Those who don't pick this up just don't develop into outstanding surgeons, frequently encountering problems, constantly fighting complications.

I first became aware of this ability when a classmate pointed it out at Yale. He and I used to play table soccer (sometimes called fussball), and, although I had never played before, almost from the first lesson I did it with speed and ease. I didn't realize it then, but it was because of this ability. When I visited Yale in early 1988, I chatted with a former classmate who is on staff there. He laughingly told me that I had been so good at the game that afterward they named several plays "Carson shots."

During my studies at medical school and the years afterward I realized the value of this skill. For me it is the most significant talent God has given me and the reason people sometimes say I have gifted hands.

———

After my first year in med school, I had a summer job as a radiology technician taking X-rays—it was the only free summer I had from then on. I enjoyed it because I learned a lot about X-rays, how they worked, and how to use the equipment. I didn't realize it at the time but subsequently this would be useful to me in research.

The medical school administration offered selected seniors opportunities as instructors, and by my senior year I was doing extremely well, receiving academic honors as well as recommendations in my clinical rotations. At one point I taught physical diagnosis to first- and second-year students. In the evenings they came over, and we practiced on each other. We learned how to listen to the sounds of our hearts and lungs, for example, and how to test reflexes. It was an incredibly good experience, and the job forced me to work hard to be ready for my students.

I didn't begin in the top of my class, however. In my first year of med school my work was only average. That's when I learned the importance of truly in-depth learning. I used to go to lectures without getting much from them, particularly when the speaker was boring. But I didn't learn much either.

For me, it paid to thoroughly study the textbooks for each course. During my second year I went to few lectures. Normally, I got out of bed around 6:00 a.m. and would go over and over the textbooks until I knew every concept and detail in them. Enterprising individuals took extremely good notes for the lectures and then, for a little cash, sold their notes. I was one of the purchasers, and I studied the notes as thoroughly as the texts.

All during my second year, I did little else but study from the time I awakened until 11:00 at night. By the time my third year rolled around, when I could work on the wards, I knew my material cold.

---

* It came as no surprise to me that during her senior year with the Yale Symphony Orchestra, Candy performed in the European premier of the modern opera *Mass* by the gifted Leonard Bernstein. She actually had a chance to meet him in Vienna.

11 ANOTHER STEP FORWARD

T*here ought to be an easier way,* I thought as I watched my instructor. A skilled neurosurgeon, he knew what he was doing, but he had difficulty locating the foramen ovale (the hole at the base of the skull). The woman on whom he was operating had a condition called trigeminal neuralgia, a painful condition of the face. "This is the hardest part," the man said as he probed with a long, thin needle. "Just locating the foramen ovale."

Then I started to argue with myself. *You're new at neurosurgery, but already you think you know everything, huh? Remember, Ben, these guys have been doing this kind of surgery for years.*

Yeah, answered another inner voice, *but that doesn't mean they know everything.*

Just leave it alone. One day you'll get your chance to change the world.

I would have stopped arguing with myself except I couldn't get away from thinking that there must be an easier way. Having to probe for the foramen ovale wasted precious surgery time and didn't help the patient either.

OK, smart man. Find it then.

And that's just what I decided to do.

I was doing my clinical year at the University of Michigan's School of Medicine and was in my neurosurgery rotation. Each of the rotations lasted a month, and it was during this period that the surgeon commented on the difficulty of finding the little hole at the base of the skull.

After arguing with myself for some time, I took advantage of the friends I had made the previous summer when I worked as a radiology technician. I went to them and explained what was worrying me. They were interested and gave me permission to come into their department and practice with the equipment.

After several days of thinking and trying different things, I hit upon a simple technique of placing two tiny metal rings on the back and front of the skull, and then aligning the rings so that the foramen ovale fell exactly between them. Using this technique, doctors could save a lot of time and energy instead of poking around inside the skull.

I had reasoned it this way: Since two points determine a line, I could put one ring on the outside surface of the skull behind the area where the foramen ovale should be. I then would put the other one on the front of the skull. By passing an X-ray beam through the skull, I could turn the head until the rings lined up. At that point, the foramen falls in between.

The procedure seemed simple and obvious—once I'd reasoned it out—but apparently no one had thought of it before. Fact is, I didn't tell anyone either. I was thinking of how to do a better job and wasn't concerned with impressing anybody or showing my instructors a new technique.

For a short time I tormented myself by asking, *Am I getting into a new realm of things that others haven't yet discovered? Or am I just thinking I've figured out a technique no one else has considered before?* Finally I decided that I had developed a method that worked for me and that was the important thing.

I started doing this procedure and, from actual surgery, saw how much easier it was. After two such surgeries, I told my neurosurgeon professors how I was doing it and then demonstrated for them. The head professor watched, shook his head slowly, and smiled. "That's fabulous, Carson."

Fortunately, the neurosurgery professors didn't resent my idea.*

From just having an interest in neurosurgery, the field soon intrigued me so much it became a compulsion. You may have noticed that I'd had that happen before. *I have to know more,* I'd find myself thinking. Everything available in print on the subject became an article I had to read. Because of my intense concentration and my driving desire to know more, without intending to I began to outshine the interns.

It was during my second rotation—my fourth year of med school—that I became aware that I knew more about neurosurgery than the interns and junior residents. While we were making our rounds, as part of the teaching procedure the professors questioned us as we examined patients. If none of the residents knew the answer, the professor would invariably turn to me. "Carson, suppose you tell them."

Fortunately, I always could, although I was still a medical student. And, quite naturally, knowing I excelled in this area produced a real emotional high. I had worked hard and pursued an in-depth knowledge, and it was paying off. And why not? If I was going to be a doctor, I was going to be the best, most-informed doctor I could possibly be!

About this time several of the interns and residents started turning over a few of their responsibilities to me. I don't think I'll ever forget the first time a resident said, "Carson, you know so much, why don't you carry the beeper and answer the calls? If you get into something you can't handle, just holler. I'll be in the lounge catching a little sleep."

He wasn't supposed to do that, of course, but he was exhausted, and I was so delighted to have the opportunity to practice and learn that I enthusiastically agreed. Before long the other residents were handing me their beepers or turning cases over to me.

Perhaps they were taking advantage of me—and in a sense they were—because the added responsibility meant longer hours and more work for me. But I loved neurosurgery and the excitement of being involved in actually performing the operations so much that I would have taken on even more if they had asked.

I'm sure the professors knew what was going on but they never mentioned it, and I certainly wasn't going to tell. I loved being a medical student. I was the first man on the line taking care of problems, and I was having the most fun I'd ever had in my life. No problems ever arose over my workload, and I maintained a good relationship with the interns and residents. Through all of these extra opportunities, I became convinced that I enjoyed this specialty more than anything else I tried.

Often as I walked through the wards I'd think, *If this is so great now while I'm still a student, it's going to be even better when I finish my residency.* Each day I went on rounds or to the lectures or operating theater. An attitude of excitement and adventure filled my thoughts because I knew I was gaining experience and information while sharpening my skills—all the things that would enable me to be a first-rate neurosurgeon.

By then I found myself in my fourth year of medical school, ready for my one year of internship and then my residency.

Professionally, I was heading in the right direction, without any question. As a kid, I had wanted to be a missionary doctor and then got caught up in psychiatry. Now and then, as part of our training, the medical students watched presentations in clinical medicine made by various specialists who talked about their particular field. The neurosurgeons impressed me the most. When they talked and showed us before-and-after pictures, they held my attention like none of the others.

"They're amazing," I'd say to myself. "Those guys can do anything."

But the first few times I looked down upon a human brain, or saw human hands working upon that center of intelligence and emotion and motion, working to help heal, I was hooked. Then realizing that my hands were steady and that I could intuitively see the effect *my* hands had on the brain, I knew I had found my calling. And so I made the choice that would become my career and my life.

All the facets of my career came together then. First, my interest in neurosurgery; second, my growing interest in the study of the brain; and third, acceptance of my God-given talent of eye-and-hand coordination—my gifted hands—that fitted me for this field. When I made my choice for neurosurgery, it seemed the most natural thing in the world.

In medical school during our clinical (or third) year we did rotation work for a month at a time, giving us an opportunity to experience each of the fields. I signed up for and received permission to do two neurosurgery rotations. Both times I received honors in my work.

Michigan had an outstanding neurosurgery program and except for a casual incident, I would have stayed at Michigan for my internship and residency. I believe residency works much better if you're in the same place you took your previous work.

One day I overheard a conversation that changed the shape of my plans. An instructor, unaware that I was nearby, commented to another about the chairman of our neurosurgery department. "He's on his way out," he said.

"You think it's that serious?" the other man asked.

"Without question. He told me so himself. Too much political strife."

That chance conversation forced me to rethink my future at the U of M. The change of personnel would severely damage the residency program. When an interim chairman comes on the scene, he's new, uncertain, and has no idea how long he'll stay. Along with that, chaos and uncertainty reign among the residents, loyalties often divide, and personnel changes occur. I didn't want to get caught up in that because I believed it could adversely affect my work and my future.

The combination of that piece of information and the fact that I'd long admired the Johns Hopkins complex made me decide to apply at Hopkins.

I had no trepidation at sending in my application for internship at Hopkins the fall of 1976 because I felt that I was as good as anybody else at that point in my training. I had made excellent grades and achieved high scores on the national board examinations. Only one problem faced me: Johns Hopkins accepted only two students a year for neurosurgery residency although they averaged 125 applications.

I sent in my application and within weeks received the marvelous news that I would be interviewed at Johns Hopkins. That didn't put me in the program, but it got me inside the door. I knew that with the competition as stiff as it was, they interviewed but few of the applicants.

———

The manner of Dr. George Udvarhelyi, head of the neurosurgery training program, put me at ease immediately. His office was large, tastefully decorated with antiques. He spoke with a soft Hungarian accent. The smoke from his pipe lent a sweet fragrance to the room. He began by asking questions, and I felt he honestly wanted to know my answers. I also sensed that he would be fair in his evaluation and recommendation.

"Tell me a little about yourself," Dr. Udvarhelyi began, looking across his desk at me.

His manner was straightforward, interested, and I relaxed. I took a deep breath and looked him in the eyes. Did I dare to be myself? *Help me, Lord,* I prayed. *If this is Your will for me, if this is the place You know I should be, help me to give the answers that will open the doors to this school.*

"Johns Hopkins is certainly my first choice," I began.

"It's also my only choice. This is the place where I want to be this fall."

Had I said that too strongly? I wondered. *Had I been too open about what I wanted?* I didn't know, but I had decided before going to Baltimore for the interview that, above all, I wanted to be myself and to be accepted or rejected by who I was and not because I successfully projected some kind of image through a super-sales job.

After he gained a few bits of information about me, Dr. Udvarhelyi's questions revolved around medicine. "Why did you choose to become a doctor?" he asked. His hands rested on his large desk.

"What aspirations do you have? What are your primary fields of interest?"

I tried to answer clearly and concisely each time. However, at some point in the conversation, Dr. Udvarhelyi made an oblique reference to a concert he had attended the night before.

"Yes, sir," I said. "I was there."

"You were?" he asked, and I saw the startled expression on his face. "Did you enjoy it?"

"Very much," I said, adding that the violin soloist had not been quite as good as I had expected.

He leaned forward, his face animated. "I thought the same thing. He was fine, technically fine, but—"

I don't remember the rest of the interview except that Dr. Udvarhelyi honed in on classical music and we talked for a long time, maybe an hour, about various composers and their different styles of music. I think he was taken aback by the fact that this Black kid from Detroit knew so much about classical music.

When the interview concluded and I left his office, I wondered if I had gotten Dr. Udvarhelyi off track and the digression would count against me. I consoled myself with the thought that he had brought up the topic and he had kept the subject foremost in our conversation.

Years later Dr. Udvarhelyi told me that he had made a strong case for my being accepted to Dr. Long, the chairman. "Ben," he said to me, "I was impressed with your grades, your honors and recommendations, and the splendid way you handled yourself in the interview." Although he didn't say it, I'm convinced that my interest in classical music was a decisive factor.

And I pleasantly remembered the hours of study during high school I had put into being able to compete on *College Bowl.* Ironically, the year I entered college, *College Bowl* went off the air. More than once I had scolded myself for wasting a lot of time learning about the arts when it would never be used or needed.

I learned something from that experience. No knowledge is ever wasted. To quote the apostle Paul: "And we know that all things work together for good to them that love God" (Romans 8:28). The love I learned for classical music helped draw Candy and me together and also helped me get into one of the best neurosurgery programs in the United States. When we work hard to acquire expertise or understanding in any field, it pays off. In this case, at least, I saw how it certainly had yielded results. I also believe that God has an overall plan for people's lives and the details get worked out along the way, even though we usually have no idea what's going on.

I was elated when I received word that I'd been accepted into the neurosurgery program at Johns Hopkins. Now I was going to get the chance for training at what I considered the greatest training hospital in the world.

Doubts concerning the field of medicine I should specialize in vanished. With confidence born of a good mother, hard work, and trust in God, I knew I was a good doctor. What I didn't know, I could learn. "I can learn to do anything that anybody else can do," I said to Candy several times.

Maybe I was a little overconfident. But I don't think I felt cocky, and certainly never superior. I recognized others' abilities as well. But in any career, whether it's that of a TV repairman, a musician, a secretary—or a surgeon—an individual must believe in himself and in his abilities. To do his best, one needs a confidence that says, "I can do anything, and if I can't do it, I know how to get help."

⎯⎯

Life was moving beautifully for me during this time. I'd been awarded a variety of honors for my clinical work at the University of Michigan, and now I was entering the last, and perhaps most important, phase of my training.

My private life was even better. Candy graduated from Yale in the spring of 1975, and we married July 6, between my second and third years of med school. Until our marriage, I lived with Curtis. Still unmarried at that time, he had received his discharge after four years of Naval service and then enrolled at the U of M to finish college.

Candy and I rented our own apartment in Ann Arbor, and she easily found a job with the state unemployment office. For the next two years she processed unemployment claims and kept our home while I finished med school.

It was exciting to move to Baltimore from the relatively small town of Ann Arbor. During our time there, Candy worked for Connecticut General Insurance Company. Because of her temporary status she found a job doing standard clerical-type work. She also briefly had a job selling vacuum cleaners, and then she got a job at Johns Hopkins as an editorial assistant for one of the chemistry professors.

For two years Candy typed for several different Johns Hopkins publications and did some editing. During that two-year period, she also took advantage of the opportunity of our being at Johns Hopkins and went back to school.

Since she was an employee of the university and married to a resident, Candy could go to school free. She continued with her course work and earned her master's degree in business. Then she went over to Mercantile Bank and Trust and started working in trust administration.

I worked hard as a resident at Johns Hopkins. One of

Ben's high school graduation. Sonya Carson, far right, with family friends.

Curtis and Ben at a teenage Christmas.

Sonya Carson holds the high school graduation photographs of her sons Ben, left, and Curtis.

113

Ben's first year at Yale.

Ben and Curtis at Ben's graduation from medical school.

Murray and B.J. welcome their new Christmas present.

One-day-old Rhoeyce with his father.

Detroit Free Press/William DeKay (5-15-88)
Ben and his wife, Candy, relax at home at the piano.

Carson serenade. A lullaby before bedtime.

The Carsons at home: Ben, Murray, Rhoeyce, Candy, and B.J.

A hemispherectomy reunion.

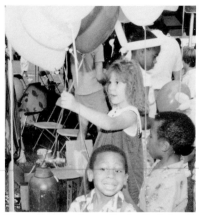

Maranda Francisco, Ben's first hemispherectomy patient, with balloons at hospital party.

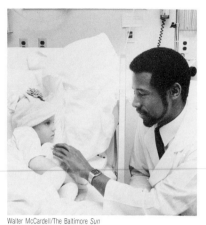

Walter McCardell/The Baltimore *Sun*

Dr. Carson talks with young patient.

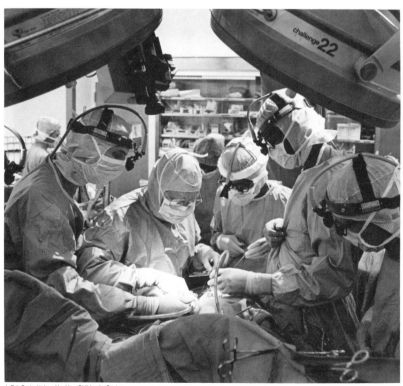

J. Pat Carter/Johns Hopkins Children's Center

The Binder twins surgery with neurosurgeons Ben Carson, Reggie Davis, Sam Hassenbusch, and Donlin Long.

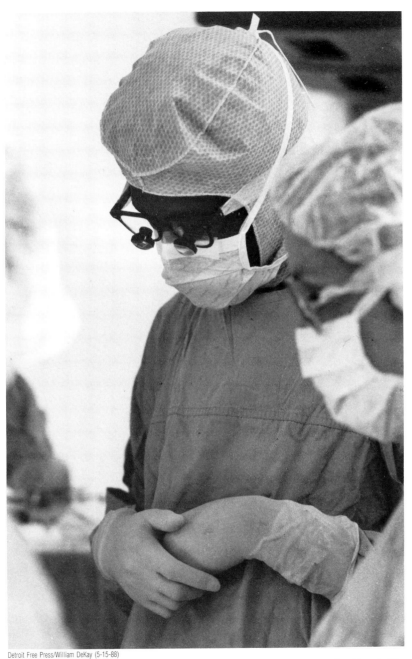

Dr. Benjamin Carson stands quietly with hands folded at Johns Hopkins Hospital in Baltimore before starting the delicate brain surgery for which he has become internationally known.

David B. Sherwin

Ben Carson receives an honorary doctoral degree from Andrews University in June 1989.

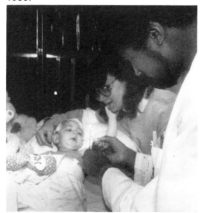

Detroit Free Press/William DeKay (5-15-88)

Dr. Carson examines 2½-year-old Megan Wikstrom during rounds at the Johns Hopkins Children's Center. "Nobody could be more patient with my one million questions and fears," says her mother, Margie Wikstrom, center.

Detroit Free Press/William DeKay (5-15-88)

Carson chats with Paul Galli, 16, of Hammonton, New Jersey, who had returned for a checkup after surgery for a brain tumor.

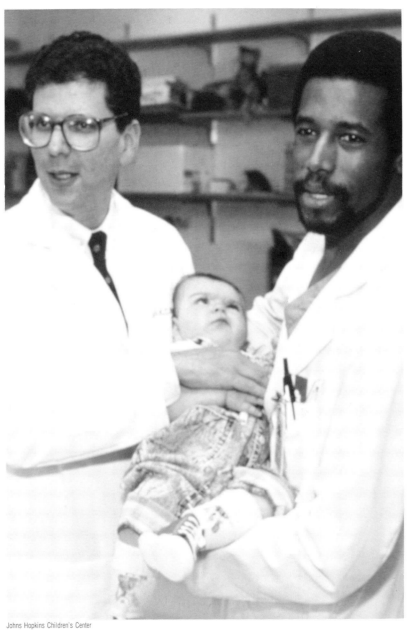

Dr. Mark Rogers and Dr. Carson with one of the Binder twins.

At a press conference after the Siamese twin surgery.

Many of the essential players of the team that separated the twins.

my goals was to maintain a good rapport with everyone because I don't believe in one-person productions. Everyone on the team is important and needs to know that he or she is vital. However, a few of the doctors tended to be snobbish, and that bothered me.

They wouldn't bother to talk with the "common people" like ward clerks or aides. That attitude troubled me, and I hurt for those dedicated employees when I saw it happening. We doctors couldn't be effective without the support of the clerks and aides. From the start I made a point to talk to the so-called lowly people and to get to know them. After all, where had I come from? I had a good teacher, my mother, who had taught me that people are just people. Their income or position in life doesn't make them better or worse than anyone else.

When I had free minutes I'd chew the fat on the wards and get to know the names of the people who worked with us. Actually this turned out to be an advantage, although I didn't plan it that way. During my residency I realized that some of the nurses and clerks had been on their jobs for 25 or 30 years. Because of their practical experience in observing and working with patients, they could teach me things. And they did.

I also realized that they recognized things that were going on with patients that I had no way of knowing. By working closely with specific patients, they sensed changes and needs before they became obvious. Once they accepted me, these often-unpraised workers quietly let me know, for instance, those I could trust or those I couldn't. They'd inform me when things were going wrong on the ward. More than once a ward clerk, on her way out the door after her shift, would pause and say, "Oh, by the way . . . " and let me know of a problem with a patient. The staff had no obligation to tell anyone, but many of them had developed an uncanny ability to sense problems, especially relapses and complications. They trusted me to listen and to act on their perceptions.

Maybe I began developing a relationship with the

staff because I wanted to compensate for the way some of the other doctors treated them. I'm not sure. I know I hated it when a resident disregarded a suggestion from a nurse. When one of them tongue-lashed a ward clerk for a simple mistake, I felt bad and a little protective toward the victim. At any rate, because of the help from the lower echelons, I was able to make an excellent showing and to do a good job.

Today I try to emphasize this point when I speak to young people. "There isn't anybody in the world who isn't worth something," I say. "If you're nice to them, they'll be nice to you. The same people you meet on the way up are the same kind of people you meet on the way down. Besides that, every person you meet is one of God's children."

I truly believe that being a successful neurosurgeon doesn't mean I'm better than anybody else. It means that I'm fortunate because God gave me the talent to do this job well. I also believe that what talents I have I need to be willing to share with others.

---

* I still use the principle of this procedure, but I've done so many of these surgeries and gotten so experienced at finding the hole, I don't need to go through the steps. I know exactly where the foramen ovale is.

12 COMING INTO MY OWN

T he nurse looked at me with disinter-
est as I walked toward her station. "Yes?" she asked,
pausing with a pencil in her hand. "Who did you come
to pick up?" From the tone of her voice I immediately
knew that she thought I was an orderly. I was wearing
my green scrubs, nothing to indicate I was a doctor.

"I didn't come to pick up anyone." I looked at her
and smiled, realizing that the only Black people she had
seen on the floor had been orderlies. Why should she
think anything else? "I'm the new intern."

"New intern? But you can't—I mean—I didn't mean
to—" the nurse stuttered, trying to apologize without
sounding prejudiced.

"That's OK," I said, letting her off the hook. It was a
natural mistake. "I'm new, so why should you know who
I am?"

The first time I went into the Intensive Care Unit, I
was wearing my whites (our monkey suits, as we interns
called them), and a nurse signaled me. "You're here for
Mr. Jordan?"

"No, ma'am, I'm not."

"You sure?" she asked as a frown covered her forehead. "He's the only one who's scheduled for respiratory therapy today."

By then I had come closer and she could read my name badge and the word *intern* under my name.

"Oh, I'm so very sorry," she said, and I could tell she was.

Although I didn't say it, I would like to have told her, "It's all right because I realize most people do things based on their past experiences. You've never encountered a Black intern before, so you assumed I was the only kind of Black male you'd seen wearing whites, a respiratory therapist." I smiled again and went on.

It was inevitable that a few White patients didn't want a Black doctor, and they protested to Dr. Long. One woman said, "I'm sorry, but I do not want a Black physician in on my case."

Dr. Long had a standard answer, given in a calm but firm voice. "There's the door. You're welcome to walk through it. But if you stay here, Dr. Carson will handle your case."

At the time people were making these objections, I didn't know about them. Only much later did Dr. Long tell me as he laughed about the prejudices of some patients. But there was no humor in his voice when he defined his position. He was adamant about his stance, allowing no prejudice because of color or ethnic background.

Of course, I knew how some individuals felt. I would have had to be pretty insensitive not to know. The way they behaved, their coldness, even without saying anything, made their feelings clear. Each time, however, I was able to remind myself they were individuals speaking for themselves and not representative of all Whites. No matter how strongly a patient felt, as soon as he voiced his objection he learned that Dr. Long would dismiss him on the spot if he said anything more. So far as I know, none of the patients ever left!

I honestly felt no great pressures. When I did encounter prejudice, I could hear Mother's voice in the back of my head saying things like, "Some people are ignorant and you have to educate them."

The only pressure I felt during my internship, and in the years since, has been a self-imposed obligation to act as a role model for Black youngsters. These young folks need to know that the way to escape their often dismal situations is contained within themselves. They can't expect other people to do it for them. Perhaps I can't do much, but I can provide one living example of someone who made it and who came from what we now call a disadvantaged background. Basically I'm no different than many of them.

As I think of Black youth, I also want to say I believe that many of our pressing racial problems will be taken care of when we who are among the minorities will stand on our own feet and refuse to look to anybody else to save us from our situations. The culture in which we live stresses looking out for number one. Without adopting such a self-centered value system, we can demand the best of ourselves while we are extending our hands to help others.

I see glimmers of hope. For example, I noticed that when the Vietnamese came to the United States they often faced prejudice from everyone—White, Black, and Hispanics. But they didn't beg for handouts and often took the lowest jobs offered. Even well-educated individuals didn't mind sweeping floors if it was a paying job.

Today many of these same Vietnamese are property owners and entrepreneurs. That's the message I try to get across to the young people. The same opportunities are there, but we can't start out as vice president of the company. Even if we landed such a position, it wouldn't do us any good anyway because we wouldn't know how to do our work. It's better to start where we can fit in and then work our way up.

My story would be incomplete if I didn't add that during my year as an intern when I was in general surgery I had a conflict with one of the chief residents, a man from Georgia named Tommy. He couldn't seem to accept having a Black intern at Johns Hopkins. He never said anything to that effect, but he continually threw caustic remarks my way, cutting me short, ignoring me, sometimes being just plain rude.

On one occasion the underlying conflict came into the open when I asked, "Why do we have to draw blood from this patient? We still have—"

"Because I said so," he thundered.

I did what he told me.

Several times that day when I asked questions, especially if they began with "Why," he snapped back the same reply.

Late that afternoon something happened that had nothing to do with me, but he was angry and, I knew from experience, would stay that way for a long time. He spun toward me, beginning, as he often did with, "I'm a nice guy, but—" It hadn't taken me long to learn that those words contradicted his nice-guy image.

This time he really laid into me. "You really do think you're something because you've had an early acceptance into the neurosurgery department, don't you? Everybody is always talking about how good you are, but I don't think you're worth salt on the earth. As a matter of fact, I think you're lousy. And I want you to know, Carson, that I could get you kicked out of neurosurgery just like that." He continued to rant for several minutes.

I just looked at him and didn't say a word. When he finally paused, I asked in my calmest voice, "Are you finished?"

"Yeah!"

"Fine," I answered calmly.

That's all I said—all that was necessary—and he stopped ranting. He never did anything to me, and I wasn't concerned about his influence anyway. Although he was the chief resident, I knew that the chiefs of the departments were the ones who made the decisions. I was determined that I wasn't going to let him make me react because then he would be able to get to me.

Instead I did my duties as I saw fit. Nobody else ever voiced any complaints about me, so I wasn't overly concerned about what he had to say.

In the general surgery department, I encountered several men who acted like the pompous, stereotyped surgeons. It bothered me and I wanted out of that whole thing. When I moved to neurosurgery it wasn't like that. Dr. Donlin Long, who has chaired the neurosurgery department at Hopkins since 1973, is the nicest guy in the world. If anybody has earned the right to be pompous, it should be him because he knows everything and everybody, and technically he is one of the best (if not the best) in the world. Yet he always has time for people and treats everyone nicely. Since the beginning, even when I was a lowly intern, I've always found him ready to answer my questions.

He is about an inch under six feet and of average build. At the time I began my internship he had salt-and-pepper hair, heavy on the pepper. Now his hair is mostly salt. He speaks with a deep voice, and people here at Hopkins are always imitating him. He knows they do it and laughs about it himself because he's got a great sense of humor. This is the man who became my mentor.

I've admired him since the first time we met. For one thing, when I came to Hopkins in 1977 there were few Blacks and none on the full-time faculty. One of the chief residents in cardiac surgery was Black, Levi Watkins, and I was one of two Black interns in general surgery, the other being Martin Goines, who had also gone to Yale. [1]

Many do their internship in general surgery but fewer in neurosurgery. Some years nobody from the Hopkins' general surgery programs division goes into neurosurgery. At the end of my intern year, five out of our group of 30 showed interest in going into neurosurgery. Of course, there were also the 125 people from other places around the country who wanted one of those slots. That year Hopkins had only one open slot.

After my year of internship I faced six years of residency, one more year of general surgery, and five of neurosurgery. I was supposed to do two years of general surgery because I applied for neurosurgery, but I didn't want to do it. I didn't like general surgery and I wanted to get out. I disliked general surgery so much I was willing to sacrifice trying for a position in the neurosurgery department at Hopkins and go somewhere else if they would take me after only one year.

I had gotten an extremely good recommendation through all my rotations as an intern. I was finishing my month rotation as an intern on the neurosurgery service and was reaching the point of writing to other schools.

However, Dr. Long called me into his office. "Ben," he said, "you've done an extremely fine job as an intern."

"Thank you," I answered, pleased to hear those words.

"Well, Ben, we've noted that you've done extremely well on your rotation on the service. All of the attendings [i.e., surgeons] have been quite impressed with your work."

Despite the fact that I wanted my features to remain passive, I know I must have been grinning widely.

"It's like this," he said and leaned slightly forward. "We'd be interested in having you join our neurosurgery program next year rather than your doing an additional year's work in general surgery."

"Thank you," I said, feeling my words were so inadequate.

His offer was a definite answer to my prayers.

———

I was a resident at Johns Hopkins from 1978 through 1982. In 1981 I was a senior resident at Baltimore City Hospital (now Francis Scott Key Medical Center), owned by Johns Hopkins.

In one memorable instance at Baltimore City, paramedics brought in a patient who had been severely beaten on the head with a baseball bat. This beating took place during the time of a meeting of the American

Association of Neurological Surgeons in Boston. Most of the faculty was away at the meeting, including the faculty person who was covering at Baltimore City Hospital. The faculty member on duty at Johns Hopkins was supposed to be covering all the hospitals.

The patient, already comatose, was deteriorating rapidly. Naturally I was quite concerned, feeling we had to do something, but I was still relatively inexperienced. Despite making phone call after phone call, I couldn't locate the faculty member. With each call, my anxiety increased. Finally, I realized that the man would die if I didn't do something—and something meant a lobectomy [2]—which I had never done before.

What should I do? I started thinking of all kinds of roadblocks such as the medical/legal ramifications of taking a patient to the OR without having an attending surgeon covering. (It was illegal to perform such a surgery without an attending surgeon present.)

What happens if I get in there and run into bleeding I can't control? I thought. *Or if I come up against another problem I don't know how to handle? If anything goes wrong, I'll have other people second-guessing my actions and asking, "Why did you do it?"*

Then I thought, *What is going to happen if I don't operate now?* I knew the obvious answer: the man would die.

The physician's assistant, Ed Rosenquist, who was on duty knew what I was going through. He said just three words to me—"Go for it."

"You're right," I answered. Once I made the decision to go ahead, a calmness came over me. I had to do the surgery, and I would do the best job I could.

Hoping I sounded confident and competent, I said to the head nurse, "Take the patient to the operating room."

Ed and I prepared for surgery. By the time the operation actually began I was perfectly calm. I opened up the man's head and removed the frontal and temporal lobes from his right side because they were swelling so greatly. It was serious surgery, and one may wonder how the man could live without that portion of his brain. The fact is that these portions of the brain are most expendable. We had no problems during surgery. The man woke up a few hours later and subsequently was perfectly normal neurologically, with no ongoing problems.

However, that episode evoked a great deal of anxiety in me. For a few days after I'd operated, I was haunted by the thought that there might be trouble. The patient could develop any number of complications and I could be censured for performing the operation. As it turned out, no one had anything negative to say. Everyone knew the man would have died if I hadn't rushed him into surgery.

A highlight for me during my residency was the research I did during my fifth year. For a long time my interest had continued to grow in the areas of brain tumors and neuro-oncology. While I wanted to stay with doing this kind of research, we didn't have the right animals in which we could implant brain tumors. By working with small animals, researchers had long established that once they obtain consistent results, they could eventually transfer their findings toward finding cures, and then they could offer help to suffering human beings. This is one of the most fruitful forms of research to find cures for our diseases.

Researchers had done a lot of work using mice, monkeys, and dogs, but they encountered problems. Dog models produced inconsistent results; monkeys were prohibitively expensive; the murines (rats and mice) were cheap enough but so small that we couldn't operate on them. Neither did they image well with CT Scans[3] and MRI[4] equipment.

To accomplish the research I wanted, I faced a triple challenge: (1) to come up with a relatively inexpensive model, (2) to find one that was consistent, and (3) to find a model large enough to be imaged and operated on.

My goal was to work with one type of animal and let that be the basis (or model) for our developmental research in brain tumors. A number of oncologists and researchers who had previously established working models counseled, "Ben, if you go ahead and begin to research brain tumors, you'd better expect to spend at least two years in the lab on the project."

When I embarked on the project I was willing to work that long or longer. But which animals should I use? While I initially started with rats, they were actually too small for our purpose. And, personally, I hate rats! Maybe they triggered too many memories of my life in Boston's tenement district. I soon realized rats did not have the qualities necessary for good research, and I began to search for a different animal.

During the next few weeks I talked to a lot of people. One fabulous thing about Johns Hopkins is that they have experts who know practically everything about their own field. I started making the rounds among the researchers asking, "What kind of animals do you use? Have you thought of any other kind?"

After a lot of questions and many observations, I hit upon the idea of using New Zealand white rabbits. They perfectly fitted my threefold criteria.

Someone at Hopkins pointed me to the research work of Dr. Jim Anderson, who was currently using New Zealand white rabbits. It was a thrill to walk into the lab there in the Blaylock Building. Inside, I saw a large open area with an X-ray machine, a surgical table off to one side, a refrigerator, an incubator, and a deep sink. Another small section housed the anesthetics. I introduced myself to Dr. Anderson and said, "I understand that you've been working with rabbits."

"Yes, I have," he answered and told me the results he'd already obtained by working with what he called VX2 to cause tumors in the liver and kidneys. Over a period of time, his research showed consistent results.

"Jim, I'm interested in developing a brain tumor model, and I wondered about using rabbits. Do you know any tumors that might grow in rabbits' brains?"

"Well," he said, thinking aloud, "VX2 might grow on the brain."

We talked a little more and then I pushed him. "Do you really think VX2 will work?"

"I don't see any reason why not. If it'll grow in other areas, there's a good chance it might grow on the brain." He paused and added, "If you want to, try it."

"I'm game."

Jim Anderson aided me immensely in my research. We first tried mechanical disassociation; that is, we used little screens to grate the tumors, much like someone would grate cheese. But they didn't grow. Second, we implanted chunks of tumors into the rabbits' brains. This time they grew.

To do what we call viability testing, I approached Dr. Michael Colvin, a biochemist in the oncology lab, and he sent me to another biochemist, Dr. John Hilton.

Hilton suggested using enzymes to dissolve the connected tissue and leave the cancer cells intact. After weeks of trying different combinations of enzymes, Hilton came up with just the right combination for us. We soon had high viability—almost 100 percent of the cells survived.

From there we concentrated the cells in the quantities we wanted. By refining the experiments we also developed a way of using a needle to implant them into the brain. Soon almost 100 percent of the tumors grew. The rabbits uniformly died with a brain tumor somewhere between the twelfth and fourteenth day, almost like clockwork.

When researchers have that kind of consistency they can go on to learn how brain tumors grow. We were able to do CT scans and became excited when the tumors actually showed up. The Magnetic Resonance Imaging (MRI), developed in West Germany, was a new technology just breaking on the scene at that time, and wasn't available to us.

Jim Anderson took several of the rabbits to Germany, imaged them on the MRI, and was able to see the tumor. I would have loved to go with him and would have, except that I didn't have the money for the trip.

Then we had the use of a PET [5] scanner in 1982. Hopkins was one of the first places in the country to get one. The first scans that we did on it were the rabbits with the brain tumors. Through the medical journals we received wide publicity for our work. To this day a lot of people at Johns Hopkins and other places are working with this brain tumor model.

Ordinarily this research would have taken years to accomplish, but I had so much collaborative effort with others at Hopkins helping to iron out our problems that the model was complete within six months.

For this research work I won the Resident of the Year Award. This also meant that instead of staying in the lab for two years I came out the next year and went on to do my chief residency.

I began my year of chief residency with a quiet excitement. It had been a long, sometimes tough road. Long, long hours, time away from Candy, study, patients, medical crises, more study, more patients—I was ready to get my hands on surgical instruments and to actually learn how to perform delicate procedures in a quick, efficient way. For example, I learned how to take out brain tumors and how to clip aneurysms. Different aneurysms require different sized clips, often put on at an odd angle. I practiced until the clipping procedure became second nature, until my eyes and instinct told me in a moment the type of clip to use.

I learned to correct malformations of bone and tissue and to operate on the spine. I learned to hold an air-powered drill, weigh it in my hand, test it, then use it to cut through bone only milimeters away from nerves and brain tissue. I learned when to be aggressive and when to hold back.

I learned to do the surgery that corrects seizures. Learned how to work near the brain stem. During that intense year as chief resident, I learned the special skills that transformed the surgical instruments along with my hands, my eyes, and intuition into healing.

Then I finished the residency. Another chapter of my life was ready to open and, as often happens before life-changing events, I wasn't aware of it. The idea came across as impossible—at first.

[1] Martin Goines is now an otolaryngologist (ear, nose, and throat) at Sinai Hospital in Baltimore and the chief of the division.

[2] Lobectomy means actually taking out the frontal lobe, while lobotomy means just cutting some fibers.

[3] Commonly called Cat Scans for Computerized Tomography, a highly technical, sophisticated computer that allows the X-ray beams to focus at different levels.

[4] The Magnetic Resonance Imaging doesn't use X-rays but a magnet that excites the protons (microparticles), and the computer then gathers energy signals from these excited protons and transforms the protons into an image.

MRI gives a clear-cut, definite picture of substances inside by reflecting the image based on the excitation of the protons. For instance, protons will be excited in a different degree in water than in bones or muscles or blood.

All protons give off different signals, and the computer then translates them into an image.

PET (Positron Emission Tomography) uses radioactive substances that can be metabolized by cells and gives off radioactive signals that can be picked up and translated. Just like the magnetic resonance imagery picks up electronic signals, this picks up radioactive signals and translates them into images.

13 A SPECIAL YEAR

I didn't explain the real reason to Bryant Stokes. I figured he knew it without my having to bring it out in the open. Instead I answered, "Sounds like a nice place." Another time I said, "Who knows? Maybe one day."

"Be a grand place for you," he persisted.

Each time he mentioned it, I gave Stokes another excuse, but I did think about what he said. One benefit especially appealed to me. "You'd get as much experience in neurosurgery there in one year as you'd get in five years anywhere else."

It seemed strange to me that Bryant Stokes persisted in the idea, but he did. A senior neurosurgeon in the United States from Perth, Western Australia, Bryant and I hit it off at once. Frequently Bryant would say, "You should come to Australia and be a senior registrar at our teaching hospital."

I tried various ways of getting him off the subject. "Thanks, but I don't think it's what I want to do." Or another time I said, "You've got to be kidding. Australia is on the other side of the world. You drill through from

Baltimore and you come out in Australia."

He laughed and said, "Or you could fly and be there in 20 hours."

I tried evasive humor. "If you're there, who needs me or anyone else?"

A matter of deep concern for me, which I naturally didn't mention, was that people had been telling me for years that Australia was worse with apartheid than South Africa. I couldn't go there because I'm Black and they had a Whites-only policy. Didn't he realize I was Black?

I dismissed the whole idea. Aside from the racial matter, from my perspective I couldn't see that going to Australia for a year of residency would add anything in terms of my career, although it would certainly be interesting.

If Bryant hadn't been so persistent, I wouldn't have given the idea any more thought. Virtually every time we talked, he'd make a casual remark such as, "You know, you'd love Australia."

I had other plans because Dr. Long, head of neuro-surgery and my mentor, had already told me that I could stay on the faculty of Johns Hopkins after my residency. The fact that he added, "I'd be delighted to have you," made it all that more appealing.

I couldn't think of anything more exciting than to remain at Hopkins, where so much research was going on. For me, Baltimore had become the center of the universe.

Yet, strange as it seemed, although I'd dismissed Australia, the topic dogged me. It seemed that for a while every time I went somewhere, I'd encounter someone with that particular accent saying, "Ga'day, mate, how you going?"

Turning on the television, I hit commercials saying, "Travel to Australia and visit the land of the koala." And PBS advertised a special on the land down under.

Finally I asked Candy, "What in the world is going on? Is God trying to tell us something?"

"I don't know," she answered, "but maybe we'd better talk a little about Australia."

Immediately I thought of a load of problems, the main one being the Whites-only policy. I asked Candy to go to the library and check out books on Australia so we could find out about the country.

The next day Candy phoned me. "I found out something about Australia you ought to know." Her voice held an uncommon excitement so I asked her to tell me right then.

"That Whites-only policy that's bothered you," she said. "Australia used to have it. They abolished that law in 1968."

I paused. What was happening here? "Maybe we ought to consider this invitation seriously," I told her. "Maybe we just ought to go to Australia."

The more we read, the more Candy and I liked the idea. Before long we were getting excited. Next we discussed Australia with friends. With few exceptions, our well-intentioned friends discouraged us. One of them asked, "Why do you want to go to a place like that?"

Another one said, "Don't you dare go to Australia. You'll be back in a week."

"You wouldn't make Candy go through that, would you?" asked another. "Why, she's had such a bad time already. It'll be worse for her down there."

I couldn't help smiling at this friend's words. His concern was our joy—and niggling worry. Candy was pregnant, and it did seem foolish to fly to the other side of the world at this time. The problem was that in 1981, while I was chief resident, Candy became pregnant with twins. Unfortunately, she miscarried in her fifth month. Now, the following year, she was pregnant again. Because of the first experience, her doctor put her on bed rest after the fourth month. She quit her job and really looked after herself.

When the question about her condition came up, Candy smiled each time but said firmly, "They do have qualified doctors in Australia, you know."

Our friends didn't realize it but we'd already decided to go, even though we didn't consciously know it ourselves. We had gone through the formal steps of making application to the Sir Charles Gardiner Hospital of Queen Elizabeth II Medical Center, the major teaching center in Western Australia and their only referral center for neurosurgery.

I received a reply within two weeks. They had accepted me. "Guess that's our answer," I said to Candy. By then she was almost more excited about our going than I was. We would leave in June 1983 and were fully committed to the venture.

We had to be fully committed because it took every dime we had to buy our tickets—one way. We wouldn't be able to come back even if we didn't like it. I would be doing one year as a senior registrar.[1]

Several reasons made the venture attractive, one of which was the money. I would be getting a good salary in Australia—a lot more money than I'd ever made before—$65,000 for the year.[2]

And we badly needed the money.

Although the racial issue was settled, Candy and I still flew to Perth with a great deal of trepidation. We didn't know what kind of reception we'd receive. We had legitimate concerns because I'd be an unknown surgeon entering a new hospital. Despite her brave talk, Candy was pregnant and the possibility of problems stayed in our minds.

But the Australians received us warmly. Our being affiliated with the Seventh-day Adventist Church opened many doors. On our first Saturday in Australia we went to church and met the pastor and several members before worship began. During the service, the pastor announced, "We have a family from the United States with us today. They're going to be here for a year." He then introduced Candy and me and encouraged the members to greet us.

And did they! When the service concluded, everybody crowded around us. Seeing that my wife was pregnant, many women asked, "What do you need?" We had brought nothing in preparation for the baby, since we were limited in the amount of luggage we could carry from the United States, and those wonderful people started bringing in bassinets, blankets, baby strollers, and diapers (which they called nappies). We were constantly receiving invitations to dinner.

People at the hospital couldn't figure out how, within two weeks of our arrival, we had met a lot of people and were receiving a constant stream of invitations.

One of my fellow residents, who had been there five months, asked, "What are you doing tonight?"

I mentioned that we were having dinner with a certain family. The resident knew that only a few days earlier a different family had taken us on a scenic trip outside Perth.

"How in the world do you know so many people?" he asked. "You've only been over here a fortnight. It took me months to meet this many people."

"We come from a large family," I said.

"You mean you have relatives here in Australia?"

"Sort of." I chuckled and then explained, "In the church, we think of ourselves as all part of God's family. That means that we think of the people where we worship as brothers and sisters—part of our family. The church people have been treating us like the family members we are."

He'd never heard such a concept before.

———

From the day we arrived, I liked Australia. Not just the people but the land and the atmosphere. Being hired as a senior registrar also meant that I got to do most of the cases. That responsibility boosted my appreciation for being in the land down under. Even Candy became really involved, as a first violinist in the Nedlands Symphony and a vocalist in a professional group.

A full month hadn't passed when an extremely difficult case came to our attention and changed the direction of my work in Perth. The senior consultant had diagnosed a young woman as having an acoustic neuroma, a tumor that grows at the base of the skull. It causes deafness and weakness of the facial muscles, eventually resulting in paralysis. This patient also suffered from frequent and extreme headaches.

The tumor was so large that, with the consultant's decision to take it out, he told the patient that he wouldn't be able to save any of her cranial nerves.

After hearing the prognosis, I asked the senior consultant, "Do you mind if I try to do this using a microscopic technique? If it works, I can possibly save the nerves."

"It is worth trying, I'm sure."

While the words were polite enough, the real flavor of his feeling came through. I knew he was saying, "You young whippersnapper, just try, and then see yourself fail." And I couldn't blame him.

The surgery took 10 straight hours without rest. Naturally, when I finished I was exhausted, but also elated. I had completely removed the tumor *and* saved her cranial nerves. The senior consultant could tell her she would likely enjoy a complete recovery.

Within a short time after her recovery, the woman became pregnant. When the baby was born, in gratitude she named the child after her consultant because she thought he had taken out her tumor and saved her cranial nerves. She didn't know that I had done the delicate work. Actually, things are done that way. In Australia, the senior registrar works under the auspices of the consultant and he, as the top man, gets the credit for successful surgery, no matter who actually performs it.

The others on the staff, of course, knew.

After that surgery, the other senior consultants suddenly showed me enormous respect. From time to time one of them would come up to me and ask, "Say, Carson, can you cover a surgery for me?"

Eager to learn and anxious for more experience, I don't recall turning down a case—which gave me a tremendous load, far more than a normal case load would provide. In less than two months in the country, I was doing two, maybe three, craniotomies a day —opening patients' heads to remove blood clots and repair aneurysms.

It takes a lot of physical stamina to do that much surgery. Surgeons spend hours on their feet at the operating table. I could handle lengthy operations because while training under Dr. Long, I had learned his philosophy and techniques, which included how to keep going, hour after tedious hour, without yielding to fatigue. I had carefully watched everything Long did and was thankful he had removed a lot of brain tumors. The Australian neurosurgeons didn't know it, but I had brain surgery down pat. The consultants increasingly gave me a freer hand than they normally would have given a senior registrar. Because I did well and was always eager for more experience, I was soon scheduling brain surgeries one on top of another. It's not quite like an assembly line because each patient is different, but I soon became the local expert in the field.

After several months, I realized that I had a special reason to thank God for leading us to Australia. In my one year there I got so much surgical experience that my skills were honed tremendously, and I felt remarkably capable and comfortable working on the brain. Before long, the wisdom of spending a year in Australia became increasingly clear to me. Where else would I have gotten such a unique opportunity for volume surgery immediately after my residency?

I did a lot of tough cases, some absolutely spectacular. And I often thanked God for the experience and the training it provided. For instance, the fire chief in Perth had an incredibly large tumor involving all the major blood vessels around the anterior part of the base of his brain. I had to operate on the man three times to get all the tumor out. The fire chief had a rocky course, but eventually he did extremely well.

One other highlight: Candy gave birth to our first son, Murray Nedlands Carson (Nedlands was the suburb where we lived), on September 12, 1983.

And then, almost before we realized it, my year was up and Candy and I were packing to return home. What would I do next? Where would I work? The chief of surgery at Provident Hospital in Baltimore contacted me soon after my return.

"Ben, you don't want to stay over there at Hopkins," he said. "You could be so much better off with us here."

Provident Hospital concentrated on medical services for Blacks. "No one is going to refer any patients to you at Hopkins," the chief of surgery said. "Why, that institution is steeped in racism. You're going to end up wasting your talents and your career in that racist institution, and you'll never go anywhere."

I nodded, thinking, *Maybe you're right.*

I listened to everything he had to say but had to make my own decision. "Thanks for your concern," I said. "I haven't been aware of prejudice toward me at Hopkins, but you may be right. Anyway, I have to find out for myself."

"You might have to go through a lot of rejection and pain to find out," he countered.

"Maybe you're right," I repeated, flattered that he wanted me to come to Provident. Yet I knew Johns Hopkins was where I wanted to be.

Then he tried another tactic. "Ben, we badly need someone here with your skills. Think of all the good you could do for Black people."

"I appreciate the offer and the interest," I told him. And I did. I didn't like disappointing him. And I didn't have the heart to tell him that I wanted to help people of all races—just people. I did say, "Let me see what happens during the next year. If things don't work out, I'll consider it."

I never contacted him again.

I'm not sure what I expected to happen when I returned from Australia to Johns Hopkins, but it was the opposite of the prediction of the other doctor. Within weeks I started getting a lot of referrals. Soon I had more patients than I knew what to do with.

After returning to Baltimore in the summer of 1984, it quickly became evident that others accepted me as a doctor competent in surgical skills. The primary reason, for which I often thanked the Lord, was that I had been blessed with more experience during my one year in Australia than many doctors get in a lifetime of medical practice.

Within months after my return, the chief of pediatric neurosurgery left to become the chairman of surgery at Brown University. By then I was already doing most of the pediatric neurosurgery anyway. Dr. Long proposed to the board that I become the new chief of pediatric neurosurgery.[3]

He told the board that, although I was only 33, I had a wide range of experience and invaluable skills. "I am fully confident that Ben Carson can do the job," he later told me he said.

Not one person on the board of that "racist institution" objected.

When Dr. Long informed me of my appointment, I was overjoyed! I also felt deeply grateful and very humbled. For days I kept saying to myself, *I can't believe this has happened.* I think I was something like a kid who'd just had a dream come true. *Look at me, here I am the chief pediatric neurosurgeon at Johns Hopkins at 33. This can't be happening to me.*

Other people couldn't believe it either. Many parents brought very sick children to our pediatric neuro-surgery unit, often traveling great distances. When I walked into the room, more than once a parent looked up and asked, "When is Dr. Carson coming?"

"He's already here," I'd answer and smile. "I'm Dr. Carson."

I got a real kick out of watching them try to contain their expression of surprise. I didn't know how much of the surprise revolved around my being Black and how much because I was so young, probably a combination of the two.

Once we got past the introductions, I would sit down with them and start talking about their child's problem. By the time I finished with the consultation, they realized I knew what I was talking about. No one ever walked out on me.

One time when I was going to do a shunt on a little

girl, her grandmother asked, "Dr. Carson, have you ever done one of these before?"

"No, not really," I said with a straight face, "but I know how to read fairly well. I own a lot of medical books, and I take most of them with me into the operating room."

She laughed self-consciously, aware of how silly her question had been.

"Actually," I joked, "I've done a thousand at least. Sometimes 300 a week." I said it with a smile, for I didn't want her to feel embarrassed.

She laughed then, realizing from the expression on my face and my tone of voice that I was still kidding her.

"Well," she said, "I guess if you are who you are, and since you have this position, you must be all right."

She didn't offend me. I knew that she passionately loved her granddaughter and wanted to be reassured that the child was in good hands. I assumed she was really saying, "You look like you haven't even gone to medical school yet." After that kind of conversation took place a few times, I became so used to the responses that I used to look forward to the reactions.

I frequently got more of a negative response from Black patients, particularly the older ones. They couldn't believe that I was chief of pediatric neurosurgery. Or if I was, that I had earned my position. At first they eyed me suspiciously, wondering if somebody had given me my position as a token expression of integration. In that case, they assumed, I probably didn't really know what I was doing. Within minutes of our consultations, however, they relaxed and the smiles on their faces told me that I had their acceptance.

Oddly enough, White patients, even the ones in whom I could clearly detect bigotry, were often easier to deal with. I could see their minds working away, and they would ultimately reason, *This guy must be incredibly good to be in this position.*

I don't face that problem nowadays because most of the patients know who I am and what I look like before they get here. But it used to be very interesting. The problem is now the opposite because I'm known in the field and too many people say, "But we have to have Dr. Carson do the surgery. We just don't want anybody else." Consequently, my operating schedule stays filled up for months in advance.

I have the prerogative of turning down patients and, of course, I must. It's necessary to say no at times because, naturally, I can't do all the surgeries. I also believe in asking other doctors if they'd be interested in doing them. I would have never learned the skills I have today if other surgeons had not been willing to let me take interesting and challenging cases.

Within a year of my appointment at Johns Hopkins I faced one of the most challenging surgeries of my life. The little girl's name was Maranda, and I had no way of knowing the influence she would have on my career. The results of her case also had a powerful effect on the medical profession's attitude toward a controversial surgical procedure.

---

[1] The position of senior registrar doesn't exist in America but lies somewhere between being a chief resident and a junior faculty member. The senior registrars run the service and work under the consultant. Following the British medical schools, Australia has what they call consultants, who are unquestionably the top men. Under this system, a doctor remains a senior registrar for many years.

A doctor can become a consultant only when the encumbent dies; the government has a fixed number of such positions.

Although they had only four consultants in Western Australia, these men were all extremely good, among the most talented surgeons I've ever seen. Each had his own area of expertise. I benefited from all their little tricks, and they aided me in developing my skills as a neurosurgeon.

[2] The salary was so attractive because I didn't have to pay exorbitant malpractice insurance. In Australia it was only $200 a year. I know a number of prominent physicians who pay $100,000 to $200,000 a year in America. The difference lies in the fact that in Australia relatively few malpractice cases arise. Australian law forbids lawyers to take malpractice cases on a contingency basis. People who want to sue have to take money from their own pockets. Consequently the only people who sue are those upon whom doctors have made the most terrible mistakes.

[3] My official title was Assistant Professor of Neurological Surgery, Direction, Division of Pediatric Neurosurgery, the Johns Hopkins University and Hospital.

14

A GIRL NAMED MARANDA

Yours is the only hospital where we've received any real hope," Terry Francisco said. She made an effort to keep her voice steady. "We've tried so many doctors and hospitals, and they end up telling us there's nothing they can do for our daughter. Please, please help us."

It had been a long and frightening three years, and as the months had blurred into years, fear turned to despair. Desperate, her daughter nearing death, Mrs. Francisco called Dr. John Freeman here at Hopkins.

In 1985 when I first came into contact with brown-haired Maranda Francisco, I could never have guessed what an influence she would have on the direction of my career: on Maranda I would perform my first hemispherectomy.*

Although born normal, Maranda Francisco had her first grand mal seizure at 18 months, a convulsion characteristic of epilepsy that we sometimes call an electrical storm in the brain. Two weeks later Maranda suffered a second grand mal seizure, and her doctor put her on anticonvulsive medication.

By her fourth birthday, the seizures were becoming more frequent. They also changed, suddenly affecting only the right side of her body. She didn't lose consciousness; the seizures were focal (half a grand mal), originating in the left side of her brain and disrupting only the right side of her body. Each seizure left Maranda weak on her right side, sometimes unable to talk normally for as long as two hours. By the time I heard about her situation, Maranda was experiencing up to 100 seizures a day, as often as three minutes apart, making the right side of her body useless. A seizure began with trembling at the right corner of her mouth. Then the rest of the right side of her face trembled, followed by the shaking of her right arm and leg, until the whole right side of her body jerked out of control and then went slack.

"She couldn't eat," her mother told us, and finally stopped letting her daughter try. The danger of choking was too great, so they started feeding her through a nasogastric tube. Although the seizures affected only her right side, Maranda was forgetting how to walk, talk, eat, and learn, and she needed constant medication. As Don Colburn of the Washington *Post* put it in a feature article, Maranda "lived her life in brief intervals between convulsions." Only during sleep was she seizure-free. As the seizures worsened, Maranda's parents took her from specialist to specialist and received varying diagnoses. More than one physician mislabeled her a mentally retarded epileptic. Each time the family went to a new doctor or clinic with hope, they left filled with disappointment. They tried medicine, diets, and, on the advice of one doctor, a cup of strong coffee twice a day.

"My daughter has been on 35 different drugs at one time or another," Terry said. "Often they'd give her so much she wouldn't recognize me."

Yet Luis and Terry Francisco refused to give up on their only child. They asked questions. They read every piece of literature they could find. Luis Francisco managed a supermarket, so they were people with only a moderate income. Yet that didn't deter them. "If there is any place on earth to get help for Maranda, we're going to find it."

In the winter of 1984 Maranda's parents finally learned the name of their daughter's condition. Dr. Thomas Reilley at the Children's Epilepsy Center at Children's Hospital in Denver, after consulting with another pediatric neurologist, suggested a possible explanation: Rasmussen's encephalitis, an extremely rare inflammation of the brain tissue. The disease progresses slowly but steadily.

If the diagnosis was correct, Reilley knew time was short. Rasmussen's progressively leads to permanent paralysis on one side of the body, mental retardation, and then death. Only brain surgery offered a possibility to save Maranda. In Denver, physicians placed the child in a barbituate coma for 17 hours in the hope that by stopping all brain activity the seizure activity might also stop. When they brought Maranda back out, immediately she started seizures again. This at least told them that the cause of her epilepsy wasn't owing to electrical misfiring in her brain but a progressive deterioration. Again, this offered more accumulated evidence of it being Rasmussen's.

Reilley arranged for Maranda's diagnosis at UCLA Medical Center, the nearest hospital with experience in treatment of Rasmussen's. A brain biopsy enabled them to reach a further confirmation of the diagnosis. The Franciscos then received the most severe blow. "It is inoperable," doctors told them. "There is nothing we can do."

That might have been the end of Maranda's story except for her tenacious parents. Terry Francisco checked on every lead she could find. As soon as she heard of anyone who was an expert in the field of seizures she made contact. When this person couldn't help her, she would say, "Do you know anyone else? Any one who might be of some help to us?" Someone finally suggested she contact Dr. John Freeman at Johns Hopkins because of his well-earned reputation in the area of seizures. By phone Terry Francisco described everything to the pediatric chief of neurology. When she finished, she heard the most encouraging words she had received in months. "Maranda sounds like she might be a good candidate for a hemispherectomy," Dr. Freeman said.

"You mean it? You think—you think you can help?" she asked, afraid to use a word like cure after so many disappointments.

"I think there's at least a good chance," he said. "Send me her records, CT scans, and anything else you have." John had been at Stanford University Hospital before hemispherectomy fell out of favor. Although he had not performed any himself, he knew of two successful hemispherectomies and was convinced that they were viable surgical options.

Hardly daring to hope, Maranda's mother copied all the records she had and mailed them that same day. When John Freeman received the material, he studied everything carefully, then came to see me. "Ben," he said, "I'd like you to take a look at this." He handed me the records, gave me a chance to study them thoroughly, and then said, "There is a procedure for a hemispherectomy that I know you've never heard of—"

"I've heard of it," I said, "but I've certainly never done one." I had heard of it only recently when, in looking up some other material, I flipped through a medical text and saw the material about hemispherectomy and skimmed it. The information didn't offer much optimism about such surgery. "I believe a hemispherectomy could save this child," Dr. Freeman told me.

"You honestly have that much confidence in the procedure?"

"I do." His eyes held mine. "Do you think you could do a hemispherectomy on this girl?" he asked. While I considered how to answer, John went on to explain the rationale behind his faith that such a surgical procedure could be done without terrible side effects.

"Sounds reasonable to me," I answered, growing

excited about having a challenge. However, I wasn't going to jump into some new kind of surgery without more information—and John Freeman wouldn't have wanted me to anyway. "Let me get hold of some of the literature and read up on it, and then I can give you a more informed answer."

Beginning that day, I read articles and papers that detailed the problems causing the high complication rate and mortality. Then I did a lot of thinking about the procedure and examined Maranda's CT scans and records. Finally I was able to say, "John, I'm not sure, but I think it's possible. Let me consider it a little more."

John and I talked and continued to study the records, and finally he phoned the Franciscos. Both of us talked with Mrs. Francisco and explained that we would consider doing a hemispherectomy. We made her no promises, and she understood that.

"You bring her for us to evaluate," I said. "Only then can we give you a definite answer."

I was eager to meet Maranda and happy when a few weeks later her parents brought her to Hopkins for further evaluation. I recall thinking how pretty she was and felt such a heaviness for the child. Maranda, then 4 years old, was from Denver, and she used to say, "I'm from Denverado."

After extensive tests, a lot of conversation with John Freeman and a few others I consulted, I was finally ready to give them my decision. Maranda's dad had flown back home to work, so I sat down with Terry Francisco. "I'm willing to attempt a hemispherectomy," I told her. "But I want you to know that I've never done one before. It's important that you understand—"

"Dr. Carson, anything—anything you can do. Everyone else has given up."

"It's a dangerous operation. Maranda may well die in the operating room." I said the words easily enough, but I also sensed how terrible they must have sounded to that mother. Yet I felt it was important to give her every negative fact. "She might have significant limitations, including severe brain damage." I kept my voice calm, not wanting to frighten her, but I also didn't want to give her false hope.

Mrs. Francisco's eyes met mine. "And if we don't agree to the surgery, what happens to Maranda?"

"She'll get worse and die."

"Then it's not much of a choice, is it? If there is a chance for her—even if a small chance—" The earnestness of her face showed clearly the emotion she had gone through in arriving at her decision. "Oh, yes, please operate."

Once they had agreed to the surgery, Terry and Luis sat down with their daughter. Terry, using a doll, showed Maranda where I would be cutting into her head, and even drew lines across the doll. "You'll also end up with a really short haircut."

Maranda giggled. She liked that idea.

Certain her daughter understood as much as she was capable of at 4 years of age, Terry said, "Honey, if you want anything special after the operation, let me know."

Maranda's brown eyes stared at her mother's face. "No more seizures."

Tears shining in her eyes, Terry embraced her daughter. She held her as if she could never let her go. "That's what we want too," she said.

On the night before surgery I walked into the pediatric playroom. Mr. and Mrs. Francisco were sitting on the edge of the play pit, a special area that the children especially enjoy. A little giraffe on wheels stood across the room. Trucks and cars were scattered around the floor. Someone had lined stuffed animals against one wall. Mrs. Francisco greeted me calmly, cheerfully. I was amazed at her calmness and the brightness in her eyes. Her serenity encouraged me to know that she was at peace and ready to accept whatever happened. Maranda played with some toys nearby.

Although I had warned them of the possible complications of surgery at the time they consented, I wanted to make certain they heard everything again. I sat on the edge of the play pit with the couple and carefully, slowly, described every phase of the surgery.

"You've obviously had some information already about what we need to do," I said, "because you talked to the pediatric neurologist. We expect the surgery to take about five hours. There's a strong possibility that Maranda could bleed to death and die on the table. There's a chance that she'll be paralyzed and never speak again. A multitude of possibilities exist of bleeding and infection and other neurological complications. On the other hand, she might do very well and never have seizures again. We don't have a crystal ball, and there's no way to know."

"Thank you for explaining," Mrs. Francisco said. "I understand."

"There is one more thing we do know," I added. "I'd like you to understand that if we don't do anything her condition will continue to worsen until you can't keep her out of an institution. And then she will die."

She nodded, too emotional to risk speaking, but I realized she had fully grasped what I said. "The risk for Maranda is compounded," I went on. "The lesion is on the left side—her dominate half of the brain." (In most right-handed people, the left hemisphere dominates speech, language, and movement on the right side of the body.) "I want to emphasize," I said, and paused, wanting to make certain they fully understood, "the major long-term risk, even if she survives surgery, is that she'll be unable to talk, or she might be permanently paralyzed on the right side. I want you to be clear about the risk you're facing."

"Dr. Carson, we know the risk," Luis said. "Whatever is going to happen is going to happen. This is our only chance, Dr. Carson. Otherwise she might as well be dead now."

As I stood to leave, I said to the parents, "And now I have a homework assignment for you. I give this to every patient and family member before surgery."

"Anything," Terry said.

"Whatever you want us to do," Luis said.

"Say your prayers. I think that really does help."

"Oh, yes, yes," Terry said and smiled.

I always tell parents that because I believe it myself. I've not yet had anybody disagree with me. While I steer away from religious discussions with patients, I like to remind them of God's loving presence. I think what little I say is enough.

I was a little anxious as I went home that night, thinking about the operation and the potential for disaster. I had talked about it with Dr. Long, who told me he had once performed a hemispherectomy. Step by step, I went over the procedure with him. Only later did I realize that I hadn't asked him if his one surgery had been successful.

So many things could go wrong with Maranda, but I had come to the conclusion years earlier that the Lord would never get me into anything He couldn't get me out of, so I wasn't going to spend an excessive amount of time worrying. I've adopted the philosophy that if somebody is going to die if we don't do something, we have nothing to lose by trying. We surely had nothing to lose with Maranda. If we didn't proceed with the hemispherectomy, death was inevitable. We were at least giving this pretty little girl a chance to live.

I finally said, "God, if Maranda dies, she dies, but we'll know that we've done the best we could for her." With that thought I had peace and went to sleep.

---

* The procedure known as *hemispherectomy* was tried as long as 50 years ago by Dr. Walter Dandy, one of the first neurosurgeons at Johns Hopkins. The three biggest names in neurosurgical history are Harvey Cushing, Walter Dandy, and A. Earl Walker, who were, consecutively, the three people in charge of neurosurgery at Hopkins dating back to the late 1800s.

Dandy tried a hemispherectomy on a patient with a tumor, and the patient died. In the 1930s and 1940s a number of people started doing the hemispherectomy. However, the side effects and mortality associated with the surgery were so great that hermispherectomy quickly fell out of favor as a viable surgical option. In the late 1950s the hemispherectomy reemerged as a possible solution for *infantile hemiplegia* associated with seizures. Skilled neurosurgeons started doing the operation again because they now had the sophisticated help of EEGs, and it seemed in a lot of patients that all the abnormal electrical activity was coming from one part of the brain. Although the results of previous hemispherectomies had been poor, surgeons felt they could now do a better job with fewer side effects. So they tried and did at least 300 of the surgeries. But again, the morbidity and mortality turned out to be high. Many patients hemorrhaged to death in the operating room. Others developed hydrocephalus or were left with severe neurological damage and either died or were rendered physically nonfunctional.

In the 1940s, however, a Montreal doctor, Theodore Rasmussen, discovered something new about the rare disease that affected Maranda. He recognized that the disease was confined to one side of the brain, affecting primarily the opposite side of the body (since the left side of the body is controlled mainly by the right side of the brain, and vice versa). It still baffles doctors why the inflammation remains in one hemisphere of the brain and doesn't spread to the other side. Rasmussen, who had long believed that the hemispherectomy was a good procedure, continued to do them when virtually everyone else had stopped.

In 1985 when I first got interested in hemispherectomy, Dr. Rasmussen was doing a diminishing number and recorded quite a few problems. I suggest two reasons for the high failure rate. First, the surgeons selected many inappropriate patients for the operation who, consequently, did not do well afterward. Second, the surgeons lacked competence or effective skills. Again the hemispherectomy fell out of favor. Experts concluded that the operation was probably worse than the disease, so it was wiser and more humane to leave such procedures alone.

Even today no one knows the cause of this disease process, and experts have suggested possible causes: the result of a stroke, a congenital abnormality, a low-grade tumor, or the more common concept, a virus. Dr. John M. Freeman, the director of pediatric neurology at Hopkins, has said, "We're not even sure whether it's caused by a virus, although it leaves footprints like a virus."

15 HEARTBREAK

In one sense, I was moving into groundbreaking surgical procedure—if I succeeded. Surgeons had recorded so few cases of full functional recovery that most doctors wouldn't consider a hemispherectomy as viable.

I was going to do my best. And I went into the surgery with two things clear. First, if I didn't operate, Maranda Francisco would worsen and die. Second, I had done everything to prepare myself for this surgery, and now I could leave the results in God's hands.

To assist me I asked Dr. Neville Knuckey, one of our chief residents, whom I had met during my year in Australia. Neville had come to Hopkins to do a fellowship, and I considered him extremely capable.

Right from the beginning of the surgery we had problems, so that instead of the expected five hours we stayed at the operating table exactly twice that long. We had to keep calling for more blood. Maranda's brain was very inflamed, and no matter where an instrument touched, she started to bleed. It was not only a lengthy operation but one of the most difficult I'd ever done.

The dramatic surgery began simply, with an incision drawn down the scalp. The assisting surgeon suctioned away blood with a hand-held tube while I cauterized small vessels. One by one, steel clips were placed on the edge of the incision to keep it open. The small operating room was cool and quiet.

Then I cut deeper through a second layer of scalp. Again small vessels were sealed shut, and a suction tube whisked away blood.

I drilled six holes, each the size of a shirt button, in Maranda's skull. The holes formed a semicircle, beginning in front of her left ear and curving up across her temple, above and down behind the ear. Each hole was filled with purified beeswax to cushion the saw. Then with an air-powered saw I connected the holes into an incision and lifted back the left side of Maranda's skull to expose the outer covering of her brain.

Her brain was swollen and abnormally hard, making the surgery more difficult. The anesthesiologist injected a drug into her IV line to reduce the swelling. Then Neville passed a thin catheter through her brain to the center of her head where it would drain off excess fluid.

Slowly, carefully, for eight tedious hours I inched away the inflamed left hemisphere of Maranda's brain. The small surgical instruments moved carefully, a millimeter at a time, coaxing tissue away from the vital blood vessels, trying not to touch or damage the other fragile parts of her brain. The large veins along the base of her brain bled profusely as I searched for the plane, the delicate line separating brain and vessels. It was not easy to manipulate the brain, to ease it away from the veins that circulated life through her small body.

Maranda lost nearly nine pints of blood during the surgery. We replaced almost twice her normal blood volume. Throughout the long hours, nurses kept Maranda's parents up-to-date on what was happening. I thought of their waiting and wondering. When my thoughts turned to God, I thanked Him for wisdom, for helping to guide my hands.

Finally we were finished. Maranda's skull was carefully sewed back in place with strong sutures. At last Neville and I stood back. The OR technician took the last instrument from my hand. I allowed myself the luxury of flexing my shoulders, rotating my head. Neville and I and the rest of our team knew we had successfully removed the left hemisphere of Maranda's brain. The "impossible" had been accomplished. *But what happens now?* I wondered.

We didn't know if the seizures would stop. We didn't know if Maranda would ever walk or talk again. We could do only do one thing—wait and see. Neville and I stepped back as the nurses lifted off the sterile sheet and the anesthesiologist unhooked and unplugged the various instruments that had recorded Maranda's vital signs. She was taken off the ventilator and began breathing on her own.

I watched her closely, searching for any purposeful movement. There was none. She moved a little when she awakened in the OR but did not respond when the nurse called her name. She did not open her eyes. *It's early,* I thought as I glanced toward Neville. *She'll wake up before long.* But would she? We had no way of knowing for certain.

The Franciscos had spent more than 10 hours in the waiting room designed for the families of surgical patients. They had resisted the suggestions to go out for a drink or to take a short walk but had stayed there praying and hoping. The rooms are cozy, decorated in soft colors, as comfortable as a waiting room can be. Magazines, books, even jigsaw puzzles, are scattered about to help pass the time. But, as one of the nurses told me later, when the morning hours stretched into afternoon, the Franciscos grew very quiet. The worry lines in their faces said it all.

I followed Maranda's gurney out of surgery. She looked small and vulnerable under the pale green sheet as the orderly wheeled her down the hall toward the pediatric intensive care unit. An IV bottle hung from a pole on the gurney. Her eyes were swollen from being under anesthesia for 10 hours. Major fluid shifts in her body had altered the working of her lymph system, causing swelling. Having the respirator tube down her throat for 10 hours had puffed her lips badly, and her face looked grotesque.

The Franciscos, alert to every sound, heard the gurney creaking down the hallway and ran to meet us. "Wait!" Terry called softly. Her eyes were red-rimmed, her face pale. She went to the gurney, bent down, and kissed her daughter.

Maranda's eyes fluttered open for a second. "I love you, Mommy and Daddy," she said.

Terry burst into joyful tears, and Luis brushed his hand across his eyes.

"She talked!" a nurse squealed. "She talked!"

I just stood there, amazed and excited, as I silently shared in that incredible moment.

We had hoped for recovery. But none of us had considered that she could be so alert so quickly. Silently I thanked God for restoring life to this beautiful little girl. Suddenly I caught my breath in amazement, as the significance of their conversation reached *my* brain.

Maranda had opened her eyes. She recognized her parents. She was talking, hearing, thinking, responding.

We had removed the left half of her brain, the dominant part that controls the speech area. Yet Maranda was talking! She was a little restless, uncomfortable on the narrow gurney, and stretched her right leg, moved her right arm— the side controlled by the half of her brain we had removed!

The news rippled down the corridor, and the whole staff, including ward clerks and aides, ran up to see with their own eyes.

"Unbelievable!"

"Isn't that great?"

I even heard a woman say, "Praise the Lord!"

———

The success of the surgery was terribly important for Maranda and her family, but it didn't occur to me that it was particularly newsworthy. While it was a break-through event, I saw it as inevitable. If I hadn't been successful, in time another neurosurgeon would have. Yet it seemed as if everybody else thought it was a big item for the news media. Reporters started coming around, calling, wanting pictures and statements. Don Colburn from the Washington *Post* interviewed me and wrote a lengthy and remarkably accurate major article, chronicling the surgery and following the family afterward. The TV program *Evening Magazine* (called *PM Magazine* in some areas) did a two-part series on hemispherectomies.

Maranda developed an infection afterward, but we quickly cleared that up with antibiotics. She continued to improve and has done extraordinarily well. Since the surgery in August 1985, Maranda Francisco has had her one wish. She has had no more seizures. However, she does lack fine motor coordination of the fingers on her right hand and walks with a slight limp. But then, she walked with a mild limp before the operation. She's taking tap dancing lessons now.

Maranda appeared on the *Phil Donahue Show*. The producers also wanted me on the show, but I turned down the invitation for several reasons. First, I'm concerned about the image I project. I don't want to become a show-business personality or be known as the celebrity doctor. Second, I'm aware of the subtlty of being called on, acknowledged, and admired on the television circuit. The danger is that if you hear how wonderful you are often enough, you begin to believe it no matter how hard you try to resist it.

Third, although I'd done my written examination for certification as a neurosurgeon, I hadn't yet taken my oral board exams. To do the oral examination, candidates sit before a board of neurosurgeons. For a full day they ask every conceivable kind of question. Common sense told me that they might not look too kindly on someone they considered a media hotdog. I considered that I had more to lose than to gain by appearing on talk shows, so I turned it down.

Fourth, I didn't want to stir up jealousies among other professionals and to have my peers say, "Oh, that's the man who thinks he's the greatest doctor in the world." This has happened to other fine doctors through media exposure.

Because he was involved, I spoke with John Freeman about these public appearances. John is older, already a full professor, and a man I highly respect. "John," I said, "there isn't anything that anybody can do to you and it doesn't matter what any jealous doctor might think about you. You've earned your reputation, and you're already highly respected. So, in light of that, why don't you go?"

John wasn't excited about making a television appearance, but he understood my reasons. "All right, Ben," he said. He appeared on the *Phil Donahue Show* and explained how the hemispherectomy worked.

Although that was my first encounter with the media, I've tended to shy away from certain types of media coverage on television, radio, and print. Each time I'm approached, I carefully look into the offer before deciding whether it's worthwhile. "What's the purpose of the interview?" That's the main question I want answered. If the bottom line is to publicize me or to provide home entertainment, I tell them I don't want to have anything to do with it.

———

Maranda manages well without the left half of her brain because of a phenomenon we call plasticity. We know that the two halves of the brain aren't as rigidly divided as we once thought. Although both have distinct functions, one side has the major responsibility for language and the other for artistic ability. But children's brains have a considerable overlap. In plasticity, functions once governed by a set of cells in the brain are taken over by another set of cells. No one understands exactly how this works.

My theory, and several others in the field agree, is that when people are born they have undifferentiated cells that haven't developed into what they are supposed to be. Or as I sometimes say, "They haven't grown up yet." If something happens to the already differentiated cells, these undefined cells still have the capacity to change and replace those that were destroyed and take over their function. As we age, these multipotential or totipotential cells differentiate more so that fewer of them remain that can change into anything else.

By the time a child reaches the age of 10 to 12, most of those potential cells have already done what they are going to do, and they no longer have the ability to switch functions to another area of the brain. That's why plasticity only works in children.

However, I don't look only at the age of the patient. I also consider the age of onset of the disease. For instance, because of her intractable seizures, I did a hemispherectomy on 21-year-old Christina Hutchins.

In Christina's case, the onset of seizures started when she was 7 and had progressed slowly. I theorized—and it turned out to be correct—that since her brain was being slowly destroyed from the age of 7, chances were that many of her functions had been transferred to other areas during the process. Although she was older than any of my other patients, we went ahead with the hemispherectomy.

Christina is now back in school with a 3.5 grade point average.

Twenty-one of the 22 patients have been females. I can't explain that fact. Theoretically, brain tumors don't occur more often in females. I think it's a fluke and that over the course of time it will even out.

Carol James, who is my physician's assistant and my right-hand person, frequently teases me by saying, "It's because women need only half of their brain to think as well as men. That's why you can do this operation on so many women."

———

I estimate that 95 percent of the children with hemispherectomies no longer have seizures. The other 5 percent have seizures only occasionally. More than 95 percent have improved intellectually after surgery because they are no longer being constantly bombarded with seizures and don't have to take a lot of medication.

I'd say that 100 percent of their parents are delighted. Of course, when the parents are delighted at the outcome, it makes us feel better too.

Hemispherectomy surgery is becoming more accepted now. Other hospitals are starting to do it. For instance, I know that by the end of 1988 surgeons at UCLA had done at least six. So far as I know, I have done more than anybody else who is actively practicing. (Dr. Rasmussen, still alive, isn't practicing medicine anymore.)

One major reason for our high success rate at Hopkins is that we have a unique situation where we work extremely well together in pediatric neurology and neurosurgery. Contrary to what I observed a few times in Australia, in our situation we don't need to depend upon a superstar. During my year down under, I noticed that some consultants weren't interested in seeing anyone else succeed; consequently, it seemed that those under them didn't always try their best.

I also praise the cooperative efforts in our pediatric intensive care unit. In fact, this togetherness permeates every aspect of our program here, including our office staff. We're friends, we work well together, we're dedicated to alleviating pain, and we're interested in each other's problems too.

We're a team, and Ben Carson is only part of that team.

———

Of all the hemispherectomies I've done, only one patient died. Since then I've done approximately 30 others. The youngest child I've given a hemispherectomy was 3-month-old Keri Joyce. The surgery was fairly routine, but she hemorrhaged afterward because of a lack of platelets in her blood. That defect affected the residual good hemisphere. Once that problem was under control, she began to recover and has had no more seizures.

The most emotionally painful experience for me was Jennifer.*

We did our initial surgery on her when she was only 5 months old.

Jennifer was having terrible seizures, and her poor mother was devastated by it all. The seizures had started within days after birth.

After doing EEGs, CT scans, MRIs, and the usual workups, we discovered that most of the abnormal activities seemed to be coming from the back part of baby Jennifer's right hemisphere. After studying everything carefully, I decided to take out only the back part.

The surgery seemed successful. She recovered quickly, and her seizure frequency diminished markedly. She started responding to our voices and growing more alert. For a while.

Then the seizures began again. On July 2, 1987, I took her into surgery and removed the rest of the right hemisphere. The operation went smoothly with no problems. Little Jennifer woke up after the operation and started moving her entire body.

The surgery with Jennifer had taken only eight hours, far less time than many others. But I think that because she was only 11 months old, the work took far more out of me than usual. When I left the operating room I was totally exhausted—and that's not normal for me.

Shoftly after Jennifer's surgery, I left for home, a drive of 35 minutes. Two miles before I reached the house, my beeper started going off. Although the cause for the emergency could have concerned half a dozen other cases, intuitively I knew that something had happened to Jennifer. "Oh, no," I groaned, "not that child."

Since I was so close, I hurried on home, rushed into the house, and called the hospital. The head nurse told me, "Shortly after you left, Jennifer arrested. They're resuscitating her now." Quickly explaining the emergency to Candy, I jumped back into my car, and made the 35-minute trip in 20 minutes.

The team was still resuscitating the infant when I got there. I joined them and we kept on, trying everything to get her back. *God, please, please don't let her die. Please.*

After an hour and a half I looked at the nurse, and her eyes said what I already knew. "She's not coming back," I said.

It took a lot of willpower not to burst into tears over the loss of that child. Immediately I turned and hurried to the room where her parents waited. Their frightened eyes locked with mine. "I'm sorry—" I said, and that's as far as I got. For the first time in my adult life I began crying in public. I felt so bad for the parents and their terrible loss. They had gone through such a roller coaster of worry, faith, despair, optimism, hope, and grief in the 11 months of little Jennifer's life.

"She was one of those children with such a fighting spirit," I heard myself telling her parents. "Why didn't she make it?" Our team had done a good job, but we sometimes face circumstances beyond medical control.

Staring at the grief etched on the faces of Jennifer's parents was a little more than I could take. Jennifer was an only child. Her mother had significant health problems herself and was being treated at the National Institutes of Health in Bethesda. Between her own problems and that of her little girl, I wondered, *Isn't this pretty close to the trials of old Job in the Bible?*

Both parents wept, and we tried to comfort each other. Dr. Patty Vining, one of the pediatric neurologists who had been with me during the operation, came into the room. She was as emotionally affected by the loss as I was. We were both trying to comfort the family while overcome by pain ourselves.

I can't remember ever feeling such a desperate loss before. The pain hurt so deeply it seemed as if everybody in the world that I loved had died at one time.

The family was devastated but, to their credit, they were understanding. I admired their courage as they went on after Jennifer's death. They had known the chances we were taking; they also knew that a hemispherectomy was the only possible way to save their daughter's life. Both parents were quite intelligent and asked many questions. They wanted to go over the records, which we opened to them. On more than one occasion, they talked to the anesthesiologist. After I had met with them a few more times, they told me they were satisfied that we had done everything possible for their little girl.

We never discovered why Jennifer died. The operation was successful. Nothing in the autopsy showed that anything had gone wrong. As sometimes happens, the cause of her death remains a mystery.

Although I continued to function, for the next several days I lived under a cloud of depression and pain. Even today when I allow myself to dwell on the death of Jennifer, it still affects me, and I can feel tears reaching toward the surface.

As a surgeon, the hardest task I have is facing parents with bad news about their child. Since I've become a parent this is even harder because now I have some inkling of how parents feel when their child is sick. I guess that's what makes it so hard. When the news is bad, there is nothing I can do or say that makes the situation better.

I know how I would feel if one of my own sons had a brain tumor. I'd feel as if I were out in the middle of the ocean sinking, pleading for somebody, anybody, to throw me a life preserver. There is a fear beyond words, beyond rational thought. Many of the parents I see come to Hopkins with that kind of despair.

Even now I'm not sure I've fully gotten over Jennifer's death. Every time a patient dies I'll probably carry an emotional scar just as people receive an emotional wound when a family member dies.

I moved beyond the depressive cloud by reminding myself that there are a lot of other people out there who need help, and it's unfair to them for me to dwell on these failures.

As I think of my own reaction, I also realize that whenever I operate and something happens that the patient doesn't do well, I feel a keen responsibility for the outcome. Probably all doctors who care deeply about their patients react that way. A few times I have tortured myself by thinking, *If I hadn't performed the surgery, it wouldn't have happened. Or if someone else had done, it perhaps the results would have been better.*

I also know I have to act rationally about these things. Often I find comfort in knowing that the patient would have died anyway and that we made a gallant attempt to save her or him. As I look back on my own history of surgery and the work we do at Hopkins, I remind myself that thousands would have died if we hadn't operated.

Some people cope with their failures easier than others. It's probably obvious from what I've told you about my need to achieve and be the best I can be that I don't handle failure well. I've said to Candy several times, "I guess the Lord knows that, so He keeps it from happening to me often."

Despite my grief over Jennifer and the days it took for me to throw it off, I don't believe in remaining emotionally detached from patients. I work with and operate on human beings, all creatures of God, people in pain who need help. I don't know how I can work on a girl's brain—how I can have her life in my hands—and yet not become involved. I feel particularly strong attachments to children who seem so defenseless and who haven't had the chance to live a full life.

---

* This is not her real name.

16 LITTLE BETH

Beth Usher fell from a swing in 1985 and received a little bump, nothing anyone worried about then. Shortly afterward that little bump caused her first minor seizure—or so they thought. What else could have been the cause? Beth, born in 1979, had been a perfectly healthy child.

A seizure is a frightening thing, especially to parents who haven't seen one before. The medical people they contacted told them there was nothing to worry about. Beth didn't *look* sick, didn't act sick, and the doctors were comforting. "This can happen after a bump to the head," they said. "The seizures will stop."

The seizures didn't stop. A month later, Beth had a second one. Her parents started to worry. Their doctor put Beth on medication to stop the seizures, and her parents relaxed. Everything would be OK now. But a few days later, Beth had another convulsion. The medication didn't stop them. Despite good medical care, the attacks came with greater frequency.

Beth's dad, Brian Usher, was the assistant football coach at the University of Connecticut. Her mother,

Kathy Usher, helped run the athletic department's fund-raising club. Brian and Kathy sought every kind of medical information, asking questions, talking to people on and off campus, determined to find some way to stop their daughter's seizures. No matter what they did, however, the seizures increased in frequency.

To her credit, Kathy Usher is a relentless researcher. One day at the library she read an article about the hemispherectomies we were doing at Johns Hopkins. That same day she phoned Dr. John Freeman. "I'd like more information about the hemispherectomies," she began. Within minutes she had poured out her sad tale about Beth.

John scheduled an appointment for them in July 1986, and they brought Beth to Baltimore. I met them that day, and we had a lengthy discussion about Beth. John and I examined her and reviewed her medical history.

At the time Beth was doing fairly well. The seizures were less frequent, down to as few as 10 a week. She was bright and vivacious, a beautiful little girl.

As I'd done with parents before, I spelled out the worst possible results, believing that when people know all the facts they can make a wiser choice.

When she had heard everything, Kathy Usher asked, "How can we go through with this? Beth seems to be getting better."

John Freeman and I understood their reluctance and did not try to force a decision. It was a terrible decision, to think of putting their bright, happy child through a radical kind of surgery. Her life was at stake. Beth was still in good shape, which made her situation unusual. When a child is at the point of death, parents have less struggle in reaching a decision. They usually end up saying something like, "She may die. By doing nothing, we'll definitely lose her. At least with surgery, she has a chance."

With Beth, however, the parents concluded, "She's doing too well. We'd better not do it."

We did nothing to force or insist upon surgery.

The Ushers returned to Connecticut with hope, indecision, and anxiety. The weeks passed, and Beth's seizures gradually increased. As they grew more frequent, she began to lose the use of part of her body.

In October 1986 the family returned to Hopkins for further tests on Beth. I saw a serious deterioration in Beth's condition in just the three-month interval. Her speech now slurred. One of the things we wanted to know was whether Beth's speech control had transferred to her good hemisphere. We tried to find out by giving the diseased hemisphere an injection to put it to sleep. Unfortunately, the entire brain went to sleep, so we couldn't determine whether the surgery would take away Beth's ability to talk.

Since their interview in July, both John and I were convinced that a hemispherectomy was the only option for Beth. After watching her condition worsen, her parents were closer to saying, "Yes, try a hemispherectomy."

At this point, John Freeman and I not only urged them to elect the surgery, but one of us said, "The sooner the better for Beth."

The poor Ushers didn't know what to do—and I understood their dilemma. At least they now had Beth alive, although she was obviously getting worse. If she came in for surgery and it was unsuccessful, she might end up in a coma, or be fully or partially paralyzed. Or she might die.

"Go home and think about this," I suggested. "Be sure of what you want to do."

"It'll soon be Thanksgiving," John said. "Enjoy the time together. Let her have Christmas at home. But," he added gently, "please, don't let it go on after that."

Beth planned to be in a Christmas play at school, and the part meant everything to her. And then after faithfully practicing her part, while she was actually on the stage, she had a seizure. She was devastated. And so were the Ushers.

That day the family decided to go through with the hemispherectomy.

In late January 1987 they brought Beth back to Johns Hopkins. The Ushers were still a little tense but said they'd decided to go through with the surgery. We went over everything that would happen. I again explained all the risks—how she might die or be paralyzed. Watching their faces, I realized they were having a struggle to face the surgery and the possible loss of their daughter. My heart went out to them.

"We have to agree," Brian Usher said at last. "We know it's her only chance."

And so a date was set. As scheduled, Beth was wheeled to an operating room and prepared for surgery. Her parents waited, hoping and praying.

The surgery went well with no complications. But Beth remained lethargic after the operation and hard to wake up. That reaction disturbed me; that night I called for a CT scan. It showed that her brain stem was swollen, which is not abnormal, and I tried to reassure her parents, "She'll probably get better over the course of a few days once the swelling goes down."

Even as I tried to comfort the Ushers, I could see from the look on their faces they didn't believe what I was saying. I couldn't blame them for thinking I was offering the old comfort routine. Had they known me better, they would have realized that I don't take that approach. I honestly expected Beth to improve.

Kathy and Brian Usher, however, were already starting to punish themselves for allowing their child to go through this drastic surgical procedure. They had reached the stage of second-guessing where they asked each other, "What if . . . ?"

They tortured themselves by going back to the day of Beth's accident and said, "If I'd have been right there with her . . ."

"If we hadn't allowed her to play on the swing . . ."

"If we hadn't agreed to this surgery, maybe she would have deteriorated, and maybe she would have died, but we still would have another year or two with her. Now we'll never have her back again."

For hours they stood by her bed in the ICU, their

eyes on her still face, watching the rise and fall of her little chest, the whir of the respirator that kept her breathing echoing in their ears.

"Beth. Beth, darling."

Finally they left, their sad eyes caressing her face.

I felt terrible. They weren't saying anything derogatory to me, never once complaining or accusing. Yet over the years, most doctors learn to grasp unstated emotions. We also understand some of what hurting relatives go through. I was hurting inside for little Beth, and I couldn't do one thing more for her. All I could do was keep her vital signs steady and wait for her brain to heal.

Both John and I remained optimistic, and we tried to reassure them by saying, "She's going to come back. Beth's just like the kids who have severe head trauma and their brain stems swell. Sometimes they're out for days, even weeks or months, but they come back."

They wanted to believe me, and I could see they were hanging on to every word of comfort Dr. Freeman or I or the nurses could give. Yet I still didn't think they believed us.

Despite the fact that John and I believed what we told Beth's parents, we couldn't be positive that Beth would wake up or that she wouldn't, finally, just slip away. We'd never been in that particular situation before. Yet we couldn't really account for Beth's condition in any other way except that the brain stem was traumatized.

The condition wasn't so severe that she couldn't bounce back. Yet the days passed, and Beth didn't bounce back. She stayed in a comatose condition for two weeks.

Daily I examined Beth and checked her records. And it became harder every day to walk into the room and face her parents. They looked at me with despair, no longer daring to hope. Time after time I had to say, "No change yet." And I meant *yet* despite what was happening.

Everybody on the staff remained supportive, constantly offering encouragement to the Ushers. They also encouraged me as I began to grow concerned. Other doctors, even nurses, would come to me and say, "It's going to be all right, Ben."

It's always inspiring when other people try to help. They knew me and, just from my silence, they figured out what troubled me. Despite their optimistic words, it was a tough time for all of us involved with Beth Usher.

Finally Beth improved slightly, enough that she didn't have to be on a respirator, but she remained comatose. We released her from ICU and sent her down to the regular floor.

The Ushers spent as much time with her as they possibly could, regularly talking to her or playing videos for her. Beth had especially liked the TV program *Mr. Rogers' Neighborhood,* so they played video tapes of Mr. Rogers. When he heard about Beth, Fred Rogers himself even came to visit Beth. He stood by her bed, touched her hand, talked to her, but her face showed no expression and she didn't wake up.

One night her dad was lying on a cot in the room, unable to sleep. It was nearly 2:00 in the morning.

"Daddy, my nose itches."

"What?" he cried, jumping out of his cot.

"My nose itches."

"Beth talked! Beth talked!" Brian Usher ran into the hallway, so excited that he didn't realize he was wearing only underpants. I doubt that anyone cared anyway. "Her nose itches!" he yelled at the nurse.

The medical staff raced after him to the room. Beth lay quietly, a smile on her face. "It does itch. A lot."

Those words were the beginning of Beth's recovery. After that she started getting better every day.*

━━━

Each of the hemispherectomies is a story in itself. For instance, I think of 13-year-old Denise Baca from New Mexico. Denise came to us in status epilepticus, meaning she was seizing constantly. Because she had been in constant seizure for two months, she had to be on a respirator. Unable to control her breathing because of the constant convulsions, Denise had undergone a tracheotomy. Now paralyzed on one side, she hadn't spoken for several months.

Denise had been a perfectly normal child a few years earlier. Her parents took her to all the New Mexico medical centers that would examine her, and then to other parts of the country. All experts concluded that her primary seizure focus was from the speech area (Brocha's area) and from the motor cortex, the two most important sections of her dominant hemisphere.

"There is nothing that can be done for her," a doctor finally told her parents.

Those might have been the final words except that a family friend read one of the articles about Maranda Francisco. Immediately she called Denise's parents. The mother, in turn, called Johns Hopkins.

"Bring Denise here, and we'll evaluate her situation," we said.

Transporting her from New Mexico to Baltimore was no easy task because Denise was on a respirator, which required a med-e-vac—a special transport system. But they made it.

After we evaluated Denise, controversy broke out here at Hopkins over whether to do a hemispherectomy. Several neurologists sincerely thought we would be crazy to attempt such an operation. They had good reasons for their opinions. Number one, Denise was too old. Number two, the seizures were coming from areas that made surgery risky, if not impossible. Number three, she was in terrible medical condition because of her seizures. Denise had aspirated, so she was having pulmonary problems as well.

One critic in particular predicted, "She'll likely die on the table just from the medical problems, much less from a hemispherectomy." He wasn't trying to be difficult but voiced his opinion out of deep and sincere concern.

Doctors Freeman, Vining, and I didn't agree. As the three people directly involved with all the hemispherectomies at Hopkins, we had had quite a bit of experience, and we were confident that we knew more about hemispherectomies than anyone else. We reasoned that, better than anyone else at Hopkins, we ought to know her chances. She would certainly die soon without surgery. Further, despite her other medical problems, she was still a viable candidate for a hemispherectomy. And, finally, we reasoned that we three ought to be the ones to determine who was a candidate.

We talked with our critic through several conferences, supporting our arguments with the evidence and experience from our background cases. We have a conference office where we invite more than just our inner circle. Over a period of days, we presented all the evidence we could and involved any of the staff at Hopkins whom we thought might have an interest in Denise's condition.

Because of the controversy, we delayed doing the operation. Normally we would have gone ahead and done it, but we faced so much opposition we took this one slowly and carefully. Our opposition deserved a fair hearing, although we insisted upon the final word.

The neurologist-critic went so far as to write a letter to the chairman of neurosurgery, with copies to the chairman of surgery, the hospital president, and a few other people. He stated that, in his medical opinion, under no circumstances should Johns Hopkins allow this operation. He then carefully explained his reasons.

Perhaps it was inevitable that bad feelings developed over Denise's case. When these issues become important it's hard to keep personal feelings out of the picture. Because I believed in the critic's sincerity and his concern for not involving Hopkins in any extraordinary and heroic ventures, I never took his arguments as personal indictments. While I was able to stay out of any personal controversy, a few of our team members and supportive friends did get heatedly involved.

Despite all arguments he brought forth, the three of us remained convinced that Denise's only chance lay in having the surgery. We had not been forbidden to do the surgery, and no one higher up had taken any action on the objection, giving us the freedom to make our decision. Yet we hesitated, not wanting to make this a personal issue, feeling that if we did, the controversy could erupt and affect the morale of the entire hospital staff.

For days I asked God to help us resolve this problem. I pondered it as I drove back and forth to work. I prayed about it as I made my rounds, and when I knelt by my bed at night. Yet I couldn't see how it would work out.

Then the issue resolved itself. Our critic left for a five-day overseas conference. While he was gone, we decided to do the operation. It seemed like a golden opportunity, and we wouldn't have any loud outcries.

I explained to Mrs. Baca as I did to others. "If we don't do anything, she's going to die. If we do something, she may die, but at least we have a chance."

"At least the operation gives her a fighting chance," her mother said.

The parents were amenable and had been all along. They understood the issue perfectly. Denise was seizing so much and deteriorating so badly, it was becoming a race against time.

After the hemispherectomy Denise remained comatose for a few days, and then she awakened. She had stopped seizing. By the time she was ready to go home, she was starting to talk. Weeks later, Denise returned to school and has progressed nicely ever since.

———

I didn't have any animosity toward the fellow who caused the opposition, because he strongly believed that surgery was the wrong thing to do. It was his prerogative to raise objections. By his objections, he thought he was looking out for the patient's best interests as well as those of the institution.

The situation with Denise taught me two things. First, it made me feel that the good Lord won't allow me to get into a situation He can't get me out of. Second, it confirmed in me that when people know their capabilities, and they know their material (or job), it doesn't matter who opposes them. Regardless of the reputation of the critics or their popularity, power, or how much they think they know, their opinions become irrelevant. I honestly never had any doubts about Denise's surgery.

In the months afterwards, although I didn't know it at the time, I would do other and more controversial surgeries. Looking back, I believe that God had used the controversy over Denise to prepare me for the steps yet ahead.

---

* In 1988 Beth's parents reported to me that she has continued to improve. She was number one in her math class.

Beth has a slight left limp. In common with other hemispherectomies, she has limited peripheral vision on one side because the visual cortex is bilateral—the one side controls vision to the other side. For some reason vision doesn't seem to transfer. The limp has been there in every case.

17 THREE SPECIAL CHILDREN

The resident flicked off his penlight and straightened up from the bedside of Bo-Bo Valentine. "Don't you think it's time to give up on this little girl?" he asked, nodding toward the 4-year-old child.

It was early Monday morning, and I was making rounds. When I came to Bo-Bo, the house officer explained her situation. "Just about the only thing she has left is pupilary response," he said. (That meant that her pupils still responded to light.) The light he shone in her eyes told him that pressure had built inside her head. The doctors had put Bo-Bo in a barbiturate coma and given her hyperventilation but still couldn't keep the pressures down.

Little Bo-Bo was another of the far-too-many children who run out into a street and are hit by a car. A Good Humor truck struck Bo-Bo. She'd lain in the ICU all weekend, comatose and with an *intracranial* pressure monitor in her skull. Her blood pressure gradually worsened, and she was losing what little function, purposeful movement, and response to stimuli she had.

Before answering the resident, I bent over Bo-Bo

and lifted her eyelids. Her pupils were fixed and dilated. "I thought you told me the pupils were still working?" I said in astonishment.

"I did," he protested. "They were working just before you came in."

"You're telling me this just happened? That her eyes just now dilated?"

"They must have!"

"Four plus emergency," I called loudly but calmly. "We've got to do something right away!" I turned to the nurse standing behind me. "Call the operating room. We're on our way."

"Four plus emergency!" she called even louder and hurried down the corridor.

Although rare, a plus four—for dire emergency— galvanizes everyone into action. The OR staff clears out a room and starts getting the instruments ready. They work with quiet efficiency, and they're quick. No one argues and no one has time to explain.

Two residents grabbed Bo-Bo's bed and half-ran down the hallway. Fortunately surgery hadn't started on the scheduled patient, so we bumped the case.

On my way to the operating room I ran into another neurosurgeon—senior to me and a man I highly respect because of his work with trauma accidents. While the staff was setting up, I explained to him what had happened and what I was going to do.

"Don't do it," he said, as he walked away from me. "You're wasting your time."

His attitude amazed me, but I didn't dwell on it. Bo-Bo Valentine was still alive. We had a chance—extremely small—but still a chance to save her life. I decided I would go ahead and do surgery anyway.

Bo Bo was gently positioned on an "egg crate," a soft, flexible pad covering the operating table, and was covered with a pale green sheet. Within minutes the nurses and anesthiologist had her ready for me to begin.

I did a craniectomy. First I opened her head and took off the front portion of her skull. The skull bone was put in a sterile solution. Then I opened up the covering of her brain—the dura. Between the two halves of the brain is an area called the falx. By splitting the falx, the two halves could communicate together and equalize the pressure between her hemispheres. Using cadaveric dura (dura from a dead person), I sewed it over her brain. This gave her brain room to swell, then heal, and still held everything inside her skull in place. Once I covered the area, I closed the scalp. The surgery took about two hours.

Bo-Bo remained comatose for the next few days. It is heartbreaking to watch parents sit by the bedside of a comatose child, and I felt for them. I could only give them hope; I couldn't promise Bo-Bo's recovery. One morning I stopped by her bed and noted that her pupils were starting to work a little bit. I recall thinking, *Maybe something positive is starting to happen*.

After two more days Bo-Bo started moving a little. Sometimes she stretched her legs or shifted her body as if trying to get more comfortable. Over the course of a week she grew alert and responsive. When it became apparent that she was going to recover, we took her back to surgery, and I replaced the portion of her skull that had been removed. Within six weeks Bo-Bo was, once again, a normal 4-year-old girl—vivacious, bouncy, and cute.

This is another instance when I'm glad I didn't listen to a critic.

———

I've actually done one craniectomy since then. Again I encountered opposition.

In the summer of 1988, we had a similar situation except that Charles,* age 10, was in worse shape. He had been hit by a car.

When the head nurse told me that Charles's pupils had become fixed and dilated, that meant we had to take action. The clinic was especially busy that day, so I sent the senior resident to explain to the mother that, in my judgment, we ought to take Charles to the operating room immediately. We would remove a portion of his brain as a last-ditch effort to save his life. "It may not work," the resident told her, "but Dr. Carson thinks it's worth a try."

The poor mother was distraught and shocked. "Absolutely not," she cried. "I can't let you do it. You will not do that to my boy! Just let him die in peace. You're not going to be playing around with my kid."

"But this way we have a chance—"

"A chance? I want more than a chance." She kept shaking her head. "Let him go." Her response was reasonable. By then Charles wasn't responding to anything.

Only three days earlier we had regretfully told her that Charles's condition was so serious that he would probably not recover, and she should come to grips with the inevitable end. Then suddenly a man stood before her, urging her to give her consent to a radical procedure. The resident could give her no assurance that Charles would recover or even be better.

After the resident returned and repeated the conversation, I went to see Charles's mother. I spent a long time explaining in detail that we weren't going to cut the boy in pieces. She still hesitated.

"Let me tell you about a similar situation we had here," I said. "She was a sweet little girl named Bo-Bo." When I finished I added, "Look, I don't know about this surgery. It may not work, but I don't see that we can give up in a situation where we still have even a glimmer of hope. Maybe it's the smallest chance of hope, but we can't just throw it away, can we? The worst thing that could happen is that Charles dies anyway."

Once she understood exactly what I would do, she said, "You mean there really is a chance? A possibility that Charles might live?"

"A chance, yes, if we do surgery. Without it, no chance whatsoever."

"In that case," she said, "of course I want you to try. I just didn't want you cutting him up when it didn't make any difference—"

Not defending ourselves that we don't do such things, I again emphasized that this was the only chance we could offer her. She signed the consent form immediately. We rushed the boy off to the operating room.

As with Bo-Bo, it involved removing a portion of the skull, cutting between the two halves of the brain, covering the swollen brain with cadaveric dura, and sewing the scalp back up.

As expected, Charles remained comatose afterward, and for a week nothing changed. More than one staff member said something to me like, "The ball game's over. We're wasting our time."

Someone presented Charles's case in our neurosurgical grand rounds. Neurosurgery grand rounds is a weekly conference attended by all neurosurgeons and residents to discuss interesting cases. Previously scheduled for an important surgery, I couldn't be present, but I was told what was said by several who had been in on the conference.

"What do you think?" the attending doctor asked one intern.

"Isn't this going a little bit beyond the call of duty?"

Another one said quite firmly, "I think it was a foolish thing to do."

Others agreed.

One of the attending neurosurgeons, familiar with the boy's condition, stated, "These types of situations never result in anything good."

Another said, "This patient has not yet recovered, and he's not going to recover. In my opinion, it's inappropriate to attempt a craniectomy."

Would they have been so vocal if I had been present? I'm not sure, yet they were speaking from their own conviction. And since seven days had passed with no change, their skepticism was understandable.

Maybe I'm just stubborn, or maybe I inwardly knew the boy still had a fighting chance. At any rate, I wasn't ready to give up.

On the eighth day a nurse noticed that Charles's eyelids were fluttering. It was the story of Bo-Bo all over again. Soon Charles started to talk, and before a month ended, we sent him to rehab. He has made great strides ever since. In the long run, we believe he's going to be fine.

Bo-Bo won't have any seizures, but Charles may. His condition was more severe, he was older, and he didn't recover as quickly as Bo-Bo. Six months after the event (when I last had contact with the family), Charles had still not fully recovered, although he is active, walking and talking, and is developing a dynamic personality. Most of all, Charles's mother clearly is thankful to have her son alive.

Another case I don't think I'll ever forget involved Detroit-born Danielle. Five months old when I first saw her, she had been born with a tumor on her head that continued growing. By the time I saw Danielle, the tumor bulged out through the skull and was the same size as her head. The tumor had actually eroded the skin, and pus drained out of it.

Friends advised her mother, "Put your baby in an institution and let her die."

"No!" she said. "This is my child. My own flesh and blood."

Danielle's mother was doing the herculean task of taking care of her. Two or three times each day she changed Danielle's dressings, trying to keep the wounds clean.

Danielle's mother called my office because she had read an article about me in the *Ladies Home Journal* in which it stated that I frequently did surgeries that nobody else would touch. She talked to my physician's assistant, Carol James.

"Ben," Carol reported later that day, "I think this one is worth looking into."

After hearing the details, I agreed. "Have the mother send me the medical records and pictures."

Less than a week later, I examined everything. I realized immediately that it was a dismal situation. The brain was abnormal, the tumor had spread all over the place, and we didn't know how the skin could be closed.

I called my friend Craig Dufresne, a superb plastic

surgeon, and together we tried to figure out a way that we could remove the tumor and close the skull again. We also consulted Dr. Peter Phillips, one of our pediatric neuro-oncologists who specializes in treating kids with brain tumors.

Together we finally devised a way that we would actually get the tumor out. Then Dr. Dufresne would swing up muscle/skin flaps from the back and try to cover the head with them. Once that had healed, Doctors Peter Phillips and Lewis Strauss would come up with a chemotherapy program to kill any remaining malignant cells.

We assumed it was going to be a tough case and would require a tremendous amount of time. We were right. The operation to remove the tumor and to sew in the muscle flaps took 19 hours. We had no concern about the time, only the results.

Dr. Dufresne and I tag-teamed the surgery. I needed almost half of the surgery hours to remove the tumor. Then Dufresne spent the next nine hours covering her skull with the muscular cutaneous skin flaps. He was able to get the skin closed over.

About halfway through the surgery, I said to Dufresne, "I think we're going to come out of this with our socks on."

He nodded, and I could tell he felt as confident as I did.

The surgery was successful. As we had anticipated, in the weeks following the removal of the tumor, Danielle had to go back to the operating room and have the flaps moved to take tension off certain areas and to improve blood circulation to the surgical site.

Initially, Danielle started to do well and responded like a normal infant. I could see the pleasure her parents took in the everyday motions of babyhood most parents can take for granted. Her tiny hand grasping one of their fingers. A little smile. Then Danielle turned the corner and started going in the wrong direction. First, she had a small respiratory problem, followed by gastrointestinal problems. After we cleared them up, her kidneys reacted. We didn't know if these other problems were related to the tumor.

Doctors and nurses in the pediatric ICU worked around the clock trying to keep Danielle's lungs and kidneys going. They were just as involved as we were.

Finally all that could be done had been done, and she died. We did an autopsy, and we found that the tumor had metastasized all over her lungs, kidneys, and gastrointestinal tract. Our surgery for the tumor in her head was a little too late. Had we gotten to her a month earlier, before things had metastasized, we might have been able to save her.

Danielle's parents and grandparents had come from Michigan and stayed in Baltimore to be near her. During the weeks of waiting and hoping for her recovery they had been extremely dedicated, understanding, and encouraging to us in everything we tried. When she died, I marveled at their maturity.

"We want it to be clear that we don't harbor any grudges over anything you folks did here at Hopkins," Danielle's parents said.

"We've just been incredibly thankful," said the grandmother, "that you were willing to undertake a case that everyone else considered impossible anyway."

Especially I remember the words of Danielle's mother. In a voice that was barely audible, she choked back her own grief and said, "We know that you're a man of God, and that the Lord has all these things in His hands. We also believe we've done everything humanly possible to save our daughter. Despite this outcome, we'll always be grateful for everything that was done here."

I share Danielle's story because not all our cases are successful. I can count on my fingers the number of bad outcomes.

---

* For the sake of privacy I have changed his name.

18 CRAIG AND SUSAN

Twenty-five to 30 people had jammed themselves into Craig Warnick's hospital room, and they were holding a prayer meeting when I walked in. All of them were taking turns asking God for a miracle when Craig went into surgery. Not only was it amazing to see so many people crowded into the room, but even more astonishing, they had all come to pray with and for Craig.

I stayed a few minutes and prayed too. As I was leaving, Craig's wife, Susan, walked to the door with me. She gave me a warm smile. "Remember what your mother said."

"I won't forget," I answered, only too aware of Mother's words, because I had once quoted them to Susan: "Bennie, if you ask the Lord for something, believing He will do it, then He will do it."

"And you remember it too," I said.

"I believe," she said. "I really do."

Even without her saying so, I could read her confidence in the outcome of surgery.

As I walked down the hallway I thought of Susan and

Craig and all that transpired in their lives. They had already gone through so much. And it wasn't anywhere near being over.

Susan Warnick is a nurse—and an excellent one—on our children's neurosurgical floor. Her husband has a disease called Von Hippel-Lindau (VHL). Individuals with this rare disease develop multiple brain tumors as well as tumors of the retina. It's a hereditary condition. Over a period of years, Craig's father had developed four brain tumors.

Craig's ordeal began in 1974 when he was a high school senior. He learned that he had developed a tumor. Few people knew of VHL and, consequently, none of the medical profession who examined Craig anticipated other tumors. I had not yet met Craig. Another neurosurgeon operated and took out the tumor.

As I continued walking down the corridor, I thought of what he had gone through the past 13 years. Then my thoughts turned toward Susan. In her own way, she had gone through as much as Craig. I admired her for being so dedicated in taking care of Craig and making sure that everything was done for him. God had sent him the perfect mate.

Susan once said that she and Craig knew from the beginning that they had a special, heaven-sent love. They met in high school when she was 14 and he was two years older. Neither ever considered anyone else as a lifetime partner. They both became Christians in high school through the ministry of Young Life. Since then they have grown in their faith and are active members of their church.

By the time Craig was 22, they had finally learned the name of his rare disease—including the likelihood of recurring tumors. And by then he had undergone lung surgery, adrenalectomy, two brain-tumor resections, and tumors of the retinas. Despite all the physical roadblocks he faced, Craig had gone on to college between his hospitalizations. After the first surgery, Craig had trouble with his balance and swallowing

—both results of the tumor. And these two symptoms never totally left him.

In 1978 Craig started vomiting and developing headaches. Both symptoms persisted with alarming regularity. Before Craig went through tests again, both he and Susan knew he had developed another tumor. However, Craig's doctor (the original physician) did not realize it was another tumor and, as the Warnicks related the story to me, the doctor dismissed their fears.

The tests, however, confirmed that the Warnicks had been correct. The doctor set up a second surgery. The night before surgery, the Baltimore neurosurgeon said to Craig's mother, "I don't think I can remove the tumor without crippling him." While wanting to know the worst possible outcome, they were devastated, feeling he offered them little hope.

The last thing that same doctor said to Susan on the evening of April 19, 1978—the night before his second surgery—was, "Tomorrow after surgery he'll be in intensive care. Right?" He started to walk away and then turned back and added, "We hope he makes it." It was one of the few times when Susan struggled with doubt over Craig's recovery.

Craig made it through the surgery, but he had a long list of complications including double vision and the inability to swallow. His lack of balance was so bad he could not even sit up. Craig was physically miserable, emotionally depressed, and ready to give up. But Susan wouldn't give up, and she refused to allow him to stop fighting. "You are going to get well," she said constantly.

A few months later, Craig was admitted to the Good Samaritan Rehabilitation Hospital. Because of a number of significant factors involved, it was a miracle for Craig to get admitted. For the next two years, Craig had some of the best physical therapy available. And he improved dramatically.

"Thank You, God," Susan, Craig, and their families prayed, offering thanks to a loving God for every sign of progress. But for Susan and Craig, improvement was not enough. "Heavenly Father," they prayed daily, "make Craig well."

Craig had a bad time recovering and faced a variety of setbacks. No longer a husky young man, Craig lost 75 pounds—making him nothing but skin stretched over a nearly six-foot frame.

Craig continued to improve, but he still had a long way to go. He learned to feed himself. Mainly because of his trouble with swallowing, he needed an hour and a half for a meal. He couldn't walk and had to be in a wheelchair. Yet during that recovery period, showing remarkable determination, he continued in college.

The faith of those two was remarkable, especially Susan's. "He's going to walk," she told people. "Craig is going to walk again."

After two years of physical therapy, with the aid of a cane, Craig walked down the aisle with Susan, and they were married on June 7, 1980. The Baltimore *Sun* wrote a big story about this loving relationship and how it had pulled Craig from the jaws of death.

Craig threw himself into his college courses and finally completed his work. He graduated in January 1981 and found a job with the Federal government, filling a handicap quota.

But it wasn't all good news. In late 1981 Craig developed tumors in his adrenal glands. In surgery the glands were removed, and he is now on medication for the rest of his life.

Shortly afterward Susan met with Dr. Neil Miller, an opthamologist at Johns Hopkins, who told her, "At least you now have a name for the disease. It's called Von Hippel-Lindau or VHL." He smiled. "It's named for the men who discovered it." He handed Susan an article about the disease.

As she started to read, Dr. Miller told her that Von Hippel-Lindau disease strikes one person in 50,000. Characteristically, VHL causes tumors in the lung, kidneys, heart, spleen, liver, adrenal glands, and pancreas.

In that instant, Susan grasped the impact this disease would have on the rest of Craig's life. She stopped reading, and her gaze met that of Dr. Miller's. Both of them were teary-eyed.

She later said, "His crying did more to comfort me than anything he could have said. I was so impressed to discover that there were people in the medical profession who felt deeply for their patients. His crying openly made me feel he understood. And that he cared."

Susan then knew the name and characteristics of the disease. That knowledge also helped her to know what they could expect in the future—more tumors. "This disease isn't going to go away. This next surgery won't be the end of it," she said, more to herself than to Dr. Miller. "We are going to have to live with this for a lifetime, aren't we?"

Tears again filled his eyes. He nodded as he said hoarsely, "At least you know what you're dealing with now."

Susan decided not to give Craig this information. Craig is quiet by nature, and at the time he was severely depressed. She thought that if he knew the bleakness of his future, this would only add to his heavy heart.

She kept the information to herself, but she was not satisfied. She had to know more. For the next 18 months Susan read, researched, and wrote to anyone whom she thought might give any additional information.

Susan claims to have one of the largest VHL libraries in the world. And I believe her! She telephoned across the United States, finding the places where they were actually doing VHL research. Over the course of Craig's illness, Susan has become highly knowledgeable about VHL and keeps abreast of medical developments.

VHL is associated with a preventable form of blindness. Because it is a dominantly inherited disease, this means that 50 percent of the offspring of persons with VHL will eventually develop it. Craig's sister, who is now 40, had a tumor when she was in her twenties. It appears she will not have any more.

When she finally told Craig about VHL, he said simply, "I knew something serious was wrong. And the tumors kept coming back."

About that time Susan remembered how much Dr. Miller's compassion had enabled her to cope. As she thought about her experience, she concluded that nurses could benefit patients by expressing their care. It was then that she decided to enter nurse's training. After graduating in 1984, Susan applied for and received a job in the pediatric neurology department at Johns Hopkins where she has remained since. To no one's surprise, Susan is an excellent nurse.

In September, 1986, Susan realized he was showing symptoms of yet another brain tumor. That's when I came into the picture: Susan asked me to take Craig as a patient.

After I agreed, we did a CT scan, and I had to tell them that it appeared that he actually had three tumors. After some preparation, I removed the tumors and, fortunately, he didn't have any surgical complications. He did, however, have endocrinological problems which required several weeks to regulate. A little while later Craig developed another tumor in the center of the brain with a cyst in it.

A gifted chief resident named Art Wong assisted me. We had a difficult operation because we had to split the corpus callosum that connects the two halves of the brain and go right down to the center to get the thing out.

The operation went well with no problems. Craig did fine post-operatively. They were praying that this would be the last surgery while knowing the statistics worked against them. Craig continued to recover —slowly but markedly.

Then in 1988 came the dreaded news: Craig had developed another tumor, this one in his brain stem. It was in the pons—an area considered inoperable. Yet someone had to try. Craig and Susan asked me to do the surgery.

"I'm sorry," I told them. "I just can't fit Craig into my operating schedule." As Susan well knew, I was already backed up with patients. Even though I believed I made the right choice, I felt terrible having to say No.

"I'd like to have you go to one of the other neurosurgeons here at Hopkins who specializes in vascular problems," I said, "because the tumors are vascular."

"We'd really like you to do it," Craig said in his quiet voice.

"If there's any way possible," Susan said. "We know how busy you are, and we understand . . ."

After a lengthy discussion and using all my persuasion, Craig did transfer to the other surgeon's care. This man considered using a new procedure, called the gamma knife. However, after talking with the Swedish inventor of the procedure, he decided it probably wouldn't work on Craig's particular type of tumor. They would have to rethink their options.

In the meantime, Craig started to deteriorate rapidly. He lost the ability to swallow, having developed such weakness in his face that it felt numb, and he started having severe headaches. On June 19, 1988, Craig had to be admitted through the hospital emergency room.

Susan called me. As I listened, I knew I couldn't stand by and let him get worse. I had to do something. I paused as I tried to sort out my emotional reaction from my professionalism. I heard myself saying, "OK, I'm going to bump somebody off the schedule. We'll get Craig into surgery."

We scheduled him for the next day, June 20, at 6:00 p.m.

Both of them became ecstatic. I don't think I've ever seen two happier people. It seemed that just knowing I would do the surgery gave them a greater sense of peace.

"It's all in God's hands," I told them.

"But we believe you let God use your hands," Craig said.

Although I had consented to do the surgery, I had to explain to Craig and Susan that this tumor and cyst were probably in the brain stem. "I can't tell for sure until I go inside and investigate," I said. "And if it's in the brain stem—" I paused, not wanting to tell them I wouldn't be able to do anything.

"We understand," Craig said.

Susan nodded.

They grasped the odds they were facing.

"But," I added, "any part of the tumor not in the brain stem, I'll take out."

"It's going to be all right," Susan said. And she meant it. It felt a little strange having the patient's wife encourage me—for me to be on the receiving end of morale boosting.

Although I agreed to the surgery, I still did not know the best course of action. I had bandied some thoughts around, and I consulted other neurosurgeons. Nobody knew what to do about this particular tumor.

"I'm going to go in there and at least investigate," I finally said. I didn't promise the Warnicks anything —how could I? They didn't seem to need any kind of extra assurance—they were more at peace than I was.

It was the late afternoon before surgery when I found all those praying people gathered in Craig's room.

It was a tough operation. The tumor had so many abnormal blood vessels coming to and going from it that I had to use a microscope to see precisely where the tumor began so I could remove it. I looked up and down the brain stem at every angle but couldn't find anything except that his brain stem was badly swollen.

I thought, *The tumor has got to be in there within the brain stem.* So I stuck needles into the brain stem. The brain stem is considered untouchable because it has so many important structures and fibers that even the slightest irritation can cause major complications. I had already suspected that the tumor might have a cyst in it. If so, if I could reach the cyst and withdraw some fluid, it would release some of the pressure on Craig's brain.

I did not find a cyst but instead provoked terrific bleeding from the sites of the needle punctures. I couldn't get anything else to come out. After eight hours, sometime around 2:30 in the morning, we closed Craig up and sent him back to the ICU. He had gone through a lot, and I assumed he'd be totally wiped out.

I was astonished when I walked into the room the

next morning. Craig behaved as if he were pre-operative. Although lying in bed, he was smiling, moving around, even making jokes.

Once past my shock, I told Susan and him that I thought this tumor was clearly in the middle of the pons—part of the brain stem.

"I'm willing to open the pons up," I said, "but I couldn't do it last night because I'd already been operating on it for eight hours, and I was tired. I probably wouldn't be thinking right. I like to make sure I have all my faculties working when venturing into no-man's-land—something I just don't want to attempt in the middle of the night."

"Do it," Craig said.

"There isn't much choice, is there?" Susan asked.

"There is at least a 50-50 chance that Craig will die right on the table," I told Susan and Craig. Those weren't easy words to say, and yet I had to tell them all of the facts, especially the unpleasant one. "And if he doesn't die, he could be paralyzed or devastated neurologically."

"We understand," Susan said. "We want you to go ahead anyway. We are praying for a miracle. We believe God is going to do it through you."

"What have we got to lose?" Craig added. "Otherwise it's death anyway."

I scheduled the surgery for a few days later.

Although I'd known Craig and Susan were both strong Christians, more than at any other time, I saw it evidenced then. They kept saying, "We want a miracle, and we believe we're going to get one. We're praying for God to give us one."

An orderly wheeled Craig to the operating room, and the procedure began. Craig lay face down on the operating table, his head held tight onto a frame so it couldn't move. Once again, doctors shaved and scrubbed his head. A nurse placed a sterile drape over Craig with the small plastic window over the surgical site. And the surgery began.

Again it was tough going. Eventually I got down to

the side of the brain stem. "I'm going to open up a little hole in the brain stem," I murmured to my staff. I took a bipolar instrument (a small electrical coagulating instrument) and opened up the brain stem. It began to bleed profusely. Every time I touched the stem, it bled. My assistant continued to suction up the blood to keep the site clear while I asked myself *What do I do now?* I prayed silently and fervently, *God, help me know what to do.*

I always pray before any of the operations, as I scrub, standing at the table before I begin. This time I was acutely conscious of praying during the entire surgery as I kept thinking, *Lord, it's up to You. You've got to do something here.* I had no idea what to try.

I paused and stared into space as I said to God, *Craig will die unless You show me what to do.* Within seconds, I knew—a kind of intuitive knowledge filled my mind. "Let me have the laser," I said to the technician.

I asked for a laser beam simply because it seemed like the most logical choice. Using the laser, cautiously, I tried opening a little hole in the brain stem. The laser enabled me to coagulate some of the bleeding vessels as I went in. At last I got a tiny hole opened with minimum bleeding and went inside. Feeling something abnormal, I teased out a little piece of it. It was probably tumorous, but it was stuck. I tugged gently, but nothing came out. Again I hesitated, not wanting to become too aggressive. I couldn't open up the hole any larger because I was right down at the brain stem.

The anesthesiologists checked their evoked potential monitors, which showed the electrical activity coming from the brain.

"The evoked potentials are gone," one of them said.

The evoked potential had died—just the way an EKG goes flat when the heart stops beating. This flatness indicated no brain waves or activity on one side of his brain—a sign of severe damage. The brain operates on electrical activity, and the activity coming through the brain stem on that side was gone although the other side remained undamaged.

"We're in here. We're going to persist," I said, not allowing myself to consider how severe the damage might be. *God, I just can't give up. Please guide my hands.* I kept at the tiny hole in the stem, my hands easing, pleading, begging, pulling gently. Finally the tumorous growth started coming out. Gently I tugged, and suddenly it all came free in one gigantic blob.

Immediately the brain stem shrunk back closer to its normal size. But while I felt pleased that I'd gotten the growth, the damage to Craig had been done. Although I tried to keep from thinking about what would happen, I knew too well. Even if Craig did survive (which was highly unlikely), he would be a "total train wreck." He would certainly be comatose and likely paralyzed. Yet I had kept on because I knew it was the right thing to do.

The surgery continued for four more hours. When we closed up, I felt terrible. Aloud I said, "Well, we did our best." I knew I had, but my words brought me no comfort.

———

The next part of the story is told by Susan, who later taped a record of Craig's story, including her experience during the first 1988 surgery that I have just described.

SUSAN WARNICK:

A lot of friends and family members came to stay with me during the surgery that night, and I was thankful for their presence. When people weren't talking to me, I spent most of the time reading my Bible. I wanted to trust God and to push away all my doubts. But the doubts were there, gnawing at me. I couldn't grasp what was happening or understand why I was falling to pieces. I had had real confidence in God for such a long time. I was so certain that we would have a miracle. Over the years, anytime Craig showed signs of discouragement I was there to motivate him, to let him know I was with him and that we could face anything together because God was in charge of our lives. I had been so strong, and now I was falling apart.

That night nothing snapped me out of my depression. I remember saying to some of the people in the room, "I've never said this before, or felt this way before either, but right this minute I feel totally defeated. Maybe God wants me to understand that enough is enough. "Maybe Craig and I can't handle this anymore. Maybe . . . maybe it's best if it ends this way."

Naturally they tried to comfort me, but I could do nothing but wait and worry.

Sometime in the middle of the night, I looked up and saw Dr. Carson coming into the waiting room where I sat with my family. He explained about the location of the tumor, the brain damage, and said something like, "As I said before, this was likely to happen. At best, Craig will probably live a few more months and then die."

Dr. Carson has a reputation of being unflappable and showing no emotion when he talks to families. He has a soft, kind voice, so quiet that many times people have to strain to hear him. Most of all, he is always so calm.

I held myself rigid as I listened to what amounted to Craig's death sentence. The more Dr. Carson told me, the more upset I became. I didn't cry, but my whole body started trembling. I was aware of this shaking and, the more I tried to control it, the more convulsive it became. *Craig is going to die* . . . Over and over that sentence rang through my head.

Dr. Carson did say that he would try to remove this tumor if Craig and I were willing to go back to surgery again. But he also told me that Craig would definitely be paralyzed on one side of his body, " . . . and there's a possibility that he will die."

For a few minutes I hardly noticed Ben Carson or heard anything. Craig was going to die—after that nothing much registered. Dr. Carson was standing in front of me, trying to comfort me, and I knew he could never find the words that would bring me peace. After 14 years of researching VHL and having it drilled into my head that if Craig ever had a tumor in his pons, he would die, I knew what was happening. My Craig. I was going to lose him. Craig was going to die.

"The tumor was in the middle of the pons," Dr. Carson repeated. At that moment I looked up and saw Dr. Benjamin Carson, the human being. Naturally he was tired, and I could see the weariness around his eyes. But it was more than that.

This isn't the way he usually looks, I thought. *Something's different about him.* Then I knew. Dr. Carson was discouraged. Defeated.

I realized that I had been so caught up with my own confusion and pain, I had only thought of Craig and me, never considering what might be going on inside Dr. Carson.

Here was a man who masked his emotions well, and yet he wasn't doing it well right then. I thought, *This man removes half of people's brains. He does surgical procedures no one else can do.* Yet I read a sadness in his face, a look of despair.

Momentarily I forgot about Craig and myself and I felt sorry for the doctor. He had tried hard, and now he was frustrated and really down.

He finished talking, turned, and walked down the hall. As I watched him, I kept saying to myself, "I feel so sorry for him."

I ran down the hall and caught up with him. I hugged him and said, "Don't feel so bad, Ben."

I went back to the room. A patient had gone home that day, and the nurses let me spend the night in the unused room. As I lay on the bed, I stared at the ceiling. I was angry—so angry

I couldn't remember feeling that much emotion at anytime before.

"God," I whispered in the semidarkness, "we've been through so much. We've seen a lot of positive things come out of all of this.

"Even though I've had moments when it was difficult for me, especially in our early years together, this is the worst. I'm mad at You, God. You're going to let Craig die and do nothing about it. If You were going to take him, why didn't You do it in 1981? Or when he had his first tumor? If You're so loving, how can You let a person like Craig go through this much only to end up dying?

"Nothing makes sense anymore. You're going to make me a widow at 30. Craig and I will never even have a child." I recalled other women who had lost their husbands telling me that having children after their husbands' death gave them purpose, a reason to live. "They at least have children! I don't have anything!"

I hurt so deeply inside, I wanted to die.

A few minutes later I went into the bathroom and saw my reflection in the mirror. I didn't recognize the face that stared back at me. It was such a weird experience, and I stared at the stranger before me.

I walked back to the bed, more miserable than ever. I felt as if my whole life had been a mistake. "Useless! That's me. All the effort, all the caring—for nothing. And how can I live without Craig? How can You expect me to go on without him?"

The venom poured out of me. I blamed God for putting me in the position of making Craig my whole world. Now God was going to take him. I cried and let my anger spew out.

Exhausted, I finally stopped talking. In a moment of quietness, God told me something. Not a voice, and yet definitely words. *Craig is not yours that you should demand to keep him. He doesn't belong to you, Susan. He is mine.*

As the truth came through to me, I realized how foolish I had been. Craig and I had surrendered our lives to Jesus Christ back in high school. Both of us belonged to God, and I had no right to try to hold on now.

Only a few days before I had been listening to a Christian radio program. The preacher told the story of Abraham taking Isaac up the mountain and of his willingness to sacrifice him—the person Abraham loved most in life.[1]

I thought of that story and said, "Yes, God. Craig is my Isaac. And, like Abraham, I want to offer him up to You."

As I lay on the neat hospital bed, a wave of peace

198

slowly washed over me, and I slept.

———

BEN CARSON:

The afternoon following the second brain-stem surgery, I was making my rounds and went in to see Craig. I couldn't believe it—he was sitting up in bed. I stared at him several seconds and then, to cover my amazement, I said, "Move your right arm."

He moved it.

"Now your left."

Again, quite normal reactions.

I asked him to move his feet and anything else I could think of. Everything was normal. I couldn't explain how he could be normal, but he was. Craig still had problems with swallowing, but everything else seemed OK.

"I guess God had something to do with this," I said.

"I guess God had everything to do with it," he answered.

The next morning we were able to remove the breathing tube.

"Going to empty me out?" Craig laughed. He was cracking jokes, having a fine time out of all of this.

"You got your miracle, Craig," I said.

"I know." His face glowed.

I was at home with my family one evening about six weeks later when the phone rang. As soon as Susan recognized my voice, without bothering to identify herself, she shouted, "Dr. Carson! You won't believe what just happened! Craig ate a whole plate of spaghetti and meatballs! He ate it all. And he swallowed everything! That was more than half an hour ago, and he's feeling great."

We talked for some time, and it felt good to know that I had been a part of their lives during one of their special moments. It made me think of how we take simple things for granted—like the ability to swallow. Only people like Craig and Susan understand how wonderful this is.[2]

[1] See Genesis, chapter 22.

[2] What's ahead for Craig? We expect Craig to get back to his preoperative state. That means that he will be highly functional. As long as I've known him, he has been neurologically impaired. He has tremors, and he still has problems with swallowing that resulted from the devastating neurological effects of the second surgery, in which he almost died.

Unfortunately, Craig will probably have other tumors. But I think the odds of one recurring in the brain stem are small. He is currently working on a MA in pastoral counseling.

19 SEPARATING THE TWINS

I wanted to kill them and myself as well," Theresa Binder said. In January 1987, during her eighth month of pregnancy, the 20-year-old woman received the terrible news—she would give birth to Siamese twins.[1]

"Oh, my God," she cried, "this can't be true! I'm not having twins! I'm having a sick, ugly,monster!" She wept almost continuously for the next three days. In her pain this mother-to-be contemplated every possible way to avoid giving birth to the twins.

Theresa first thought of overdosing on sleeping pills to kill the unborn twins and herself. "I just couldn't go on and, for a while, it seemed like the only solution for them and for me." But when she actually faced this answer, she couldn't bring herself to swallow the pills. Some of her thoughts bordered on the bizarre, contemplating something, anything, just to have peace and to get herself out of this nightmare. She had considered running away, jumping out of the window of a tall building. No matter what she contemplated, she heard herself saying, "I just want to die."

On the fourth morning Theresa suddenly realized that she could kill herself—that would be bad enough—but that by her suicide she was murdering two other beings who had the right to live.

Theresa Binder made peace with herself, knowing that she would have to face whatever happened. Now she could move beyond the tragedy and live with the results. Other parents had.

Yet, only months before, Theresa and her 36-year-old husband, Josef, were overjoyed at the prospect of a baby. Early in her pregnancy her doctor informed them that she was carrying twins. "I was filled with joy," Theresa recalled, "and thanked God for this wonderful double gift."

In anticipation, this couple in Ulm, West Germany, bought identical baby clothes, a double cradle, and a double baby carriage as they awaited the twins' arrival.

The twins, Patrick and Benjamin, were born by Cesarean section on February 2, 1987. Together they weighed a total of eight pounds fourteen ounces, and they were joined at the back of the head.

Immediately after birth the twins were taken to the children's hospital, and Theresa didn't see them until three days later. When she finally saw her babies, Josef stood at her side, ready to catch her and carry her from the room if necessary.

She stared at the joined infants in front of her. Words like *monster* fled from her, and Theresa saw only two tiny boys—her babies—and her heart melted. Tears streamed down her face. Her husband embraced her, and then they hugged their sons. "You are ours," she said to the boys, "and I already love you."

Mother love never deserted Theresa Binder, although the days ahead were difficult—heartbreaking at times. Her protective care grew stronger.

The parents had to learn how to hold the babies to be sure they were both well supported. Because their heads turned away from each other, Theresa had to sit them against a cushion and hold a bottle of milk in each hand to feed them. Although the twins shared no vital organs, they did share a section of the skull and skin tissue, as well as a major vein responsible for draining blood from the brain and returning it to the heart.

Five weeks after their birth, the Binders took their sons home. "Not once did we ever not love them," Josef said. "They were our sons."

Because of their being joined at the heads, the boys couldn't learn to move like other infants, and yet, from the beginning, they acted like two individuals. One often slept as the other cried.

The Binders lived with the hope that their chubby, blond sons would one day be separated. As they considered the future of Patrick and Benjamin, they learned that if the boys remained attached they would never sit, crawl, turn over, or walk. The two beautiful children would remain bedridden and relegated to lying on their backs for as long as they lived. Not much of a prospect for them.

"I have lived with a dream that has kept me going," Theresa told me when we first met. "A dream that somehow we would find doctors able to perform a miracle."

Night after night as Theresa went to bed, her last thoughts centered on cuddling and holding each of her sons separately, playing with them one at a time, and putting them in different cradles. Many of those nights she lay in bed, her eyes wet with tears, wondering if there would ever be a miracle for her sons. No one had successfully separated Siamese twins joined at the back of the cranium with both surviving.[2]

"But I didn't give up hope. I couldn't. These were my sons, and they were the most important thing in my life," she said. "I knew I would fight for their chance as long as I lived."

The babies' physicians in West Germany contacted us at Johns Hopkins, asking if the pediatric surgical team could devise a plan to separate the Binder twins and give them their chance to live normal, separated lives.

That's when I came into this story.

After studying the available information, I tentatively agreed to do the surgery, knowing it would be the riskiest and most demanding thing I had ever done. But I also knew that it would give the boys a chance—their only chance—to live normally. My making that decision was only one phase, because this would not be a one-doctor procedure. Doctor Mark Rogers, Director of Pediatric Intensive Care at Hopkins, coordinated the massive undertaking. We assembled seven pediatric anesthesiologists, five neurosurgeons, two cardiac surgeons, five plastic surgeons, and, just as important, dozens of nurses and technicians—seventy of us in all. We would also undergo five months of intensive study and training-preparation for this unique surgery.

Craig Dufresne, Mark Rogers, David Nichols, and I planned to fly to West Germany in May 1987. During our four days there, Dufresne would insert inflatable silicone balloons under the scalps of the babies. This device would gradually stretch the skin so that enough tissue would be available to close the huge surgical wounds following the separation.

When it came to the surgery, I would do the actual separating, and then Donlin Long would work on one boy while I took the other. To make our chances for success better, I'd have the best qualified medical team at my side, all from Johns Hopkins, and they included Bruce Reitz, Director of Cardiac Surgery; Craig Dufresne, Assistant Professor of Plastic Surgery; David Nichols, Pediatric Anesthesiologist; and Donlin Long, chairman of Neurosurgery; with Mark Rogers as coordinator and spokesman.

Since I had seen only X-rays of the children, I needed personally to assess their neurological ability, so I would be part of the team going to Germany to determine if the surgery was still feasible.

Then two weeks before the four of us were scheduled to go, thieves broke into our house. Aside from things like electronic equipment, they also stole our safe, which they couldn't get open. The small safe, not much larger than a shoebox, contained all our important documents and papers, including our passports.

While realizing it would be difficult to replace the

passport in two weeks, I didn't know it would be impossible. When I called the state department, the kind-but-efficient voice said, "I'm sorry, Dr. Carson, but nothing can be done in such a short period."

I then asked the police investigator, "What are the chances of getting back my papers, especially the passport?"

"No chance," he snorted. "You don't ever get those kinds of things back. They trash them."

After hanging up, I prayed, "Lord, somehow You've got to get me a passport if You want me involved in this surgery." I tried not to think about the passport. Because of my caseload I became so absorbed in other things, I put the matter out of my mind.

Two days later the same policeman phoned my office. "You won't believe this, but we have your papers. And your passport."

"Oh, I believe it," I said.

In an amazed tone, he told me that a detective had been rummaging through garbage. In a big plastic bag, he found a paper with my name on it and started digging further. Then he found all the other things, every single important stolen document. From that discovery they were able to bust a large crime ring in the Baltimore-Washington, D.C., area and to recover all of our other equipment, along with items stolen from other families.

Our team spent the next five months in planning and working through every contingency we could envision. Part of the preparation required the rewiring of an entire section of a large operating room with emergency power ready in case of power failure. The OR had two of everything—anesthesia monitors, heart-lung machines, and tables that would lie side by side, but that we could move apart once I made the incision that separated the boys.

At the end of the five-month period, everything was so organized that at times it felt as if we were planning a military operation. We even worked out where each team member would stand on the operating room floor. A 10-page, play-by-play book detailed each step of the operation. We endlessly discussed the five 3-hour dress rehearsals we'd had, using life-sized dolls attached at the head by Velcro.

From the time we started discussing it, we all tried to keep in mind that we wouldn't proceed with surgery unless we believed we had a good chance of separating the boys without damaging the neurological function of either baby.

Neither Donlin Long nor I could be certain that parts of the critical brain tissue, such as the vision center, were wholly separate. Fortunately, as we had expected, the boys shared only a main drainage system, called the superior saggital sinus, a critically important vein.

———

Surgery on the 7-month-old twins began on Labor Day weekend, Saturday, September 5, 1987, at 7:15 a.m. We chose that day because the hospital itself would be less busy with plenty of staff available. (We don't schedule elective surgery on weekends.)

Mark Rogers had advised the parents to stay in their hotel room during the operation so they could get some rest. As I would have expected, they rested very little, and one of them was sitting next to the phone at all times. During the next 22 hours, one of the doctors called the Binders to update them at each stage of the ordeal.

Heart surgeons Reitz and Cameron, after anesthetizing the twins, inserted hair-thin catheters in major veins and arteries to monitor the boys during the operation. With the children's heads positioned to prevent them from sagging and causing undue pressure on the skulls after separation, we cut into the scalp and removed the bony tissue that held the two skulls, carefully preserving it so that we could use it later to reconstruct their skulls.

Next, we opened the dura—the covering of the brain. This was quite complex because of a number of convolutions or tortuous areas in the dura and in the dural plains between their brains, as well as a large, abnormal artery running between the two brains which had to be sectioned.

We had to complete all the sectioning of adhesions between the two brains before we made any attempt to separate the large venus sinuses. We isolated the top portion of the sinus and the bottom portion just below the torqula, the place where all the sinuses come together. Normally this ranges in size from that of a quarter to a half-dollar. Unfortunately it was much larger.

When we cut below the area where the torqula should have ended, we encountered fierce bleeding. We controlled the bleeding by sewing muscle patches into the area, but it was frightening bleeding. We proceeded further down, and I recall saying aloud, "The torqula can't extend much further." Yet each time we met with the same scenario. Eventually we got all the way to the base of the skull where the spinal cord and the brain stem meet, and we were still having the same problem.

We concluded that the torqula, instead of being the size of a half-dollar, covered the entirety of the backs of both of their heads and was a gigantic, highly pressurized, venus lake.

This situation forced us to go into hypothermic arrest prematurely. In the planning sessions we had carefully timed it to take from three to five minutes to separate the vascular structures and the remaining time simultaneously reconstructing them in both infants.

We had each child hooked up to a heart-lung bypass machine and pumped their blood through it to cool their temperatures from 95 degrees Fahrenheit to 68 degrees.

Slowly we removed blood from the boys' bodies. This deep degree of hypothermia brings metabolic functions to a near halt, and allowed us to stop the heart and blood flow for approximately an hour without causing brain damage. We had to stop the blood flow long enough to construct separate veins. During this time the Binder twins remained in a state similar to suspended animation.

We had figured that after an hour the tissues' demand for nourishment supplied by the blood would cause irreparable tissue damage. This meant that once we had lowered the boys' body temperatures, we had to work quickly. (Interestingly, this technique can only be used in infants under 18 months when the brain is still developing and is flexible enough to recover from such a shock.)

Just before 11:30 p.m., 20 minutes after we started lowering their body temperatures, came the critical moment. With the skulls already open, I prepared to sever the thin blue main vein in the back of the twins' heads that carried blood out of the brain. It was the last link remaining between the little boys. That completed, we pulled the hinged table apart, and Long had one boy and I had the other. For the first time in their young lives, Patrick and Benjamin were living apart from each other.

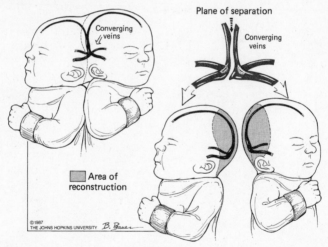

Although free, the twins immediately faced a potentially deadly obstacle. Before we could restore the blood flow, working as two units, both Long and I would have to fashion a new satittal vein from the pieces of pericardium (the covering of the heart) removed earlier.

Someone started the big timer on the wall. We had one hour to complete our work and to restart the blood flow. We were racing against time, but I said to the nursing staff, "Please don't tell me what time it is or how much time we have left." We didn't want to know; we didn't need the extra pressure of someone saying "You've only got 17 minutes left." We were working as fast as we could.

I had instructed them, "When the hour is up, just turn the pumps back on. If they bleed to death then they'll have to bleed to death, but we'll know we did the best we could." Not that I felt so heartless, but I didn't want to take the chance of brain damage. Fortunately both Long and I were used to working under pressure, and we stayed at it, not letting our attention waver.

It was an eerie experience, starting the surgery, because their bodies were so cold it was like working on a cadaver. In one sense the twins were dead. Momentarily I wondered if they would ever live again.

In the planning sessions I had anticipated that it would take about three to five minutes to cut through the sinuses. Then we would spend the remaining 50-55 minutes reconstructing the sinuses before we could turn the blood back on.

"Oh, no," I mumbled under my breath—I had hit a snag. I would need more time than I had planned to reconstruct the huge torqula on my twin. The torqula is the dreaded area for neurosurgeons because the blood rushes through that area under such pressure that a hole in the torqula the size of a pencil would cause a baby to bleed to death in less than a minute.

After hypothermic arrest it took 20 minutes to separate all of the vascular tissue, which meant we had used at least three times as much time as we had planned.

We hadn't been able to predetermine this situation because the pressure in this vascular lake was so high that it washed out the dye during the angiogram.

By using 20 minutes to separate the vessels, this gave us only 40 minutes to complete our work. Fortunately the cardiovascular surgeons had been looking over our shoulders and observing the configuration of the sinuses as I was cutting through them. From pericardium they cut pieces to exactly the right diameter and shape.

Although they were estimating, these two men were so skillful that when they handed the pericardium to Long and me, all of the pieces fitted perfectly. We were able to sew them into place along the affected areas.

At one point, perhaps 45 minutes into the hour, I knew we were moving close to the deadline. Without looking around I sensed the tension level around me increasing, almost as if individuals were whispering to each other, "Are we going to finish in time?"

Long completed his baby first. I completed mine within seconds before the blood started to flow again. We were right on target.

A silence momentarily filled the operating room, and I was conscious only of the rhythmic humming of the heart-lung machine.

"It's done," somebody said behind me.

I nodded, exhaling deeply, suddenly aware that I'd been holding my breath during those last critical moments. The strain was telling on all of us, but we had refused to give in to it.

Once we restarted the infants' hearts, we hit our second big obstacle, profuse bleeding from all the tiny blood vessels in the brain that had been severed during surgery.

Everything that could bleed did bleed. We spent the next three hours using everything known to the human mind to get the bleeding controlled. At one point we were certain we wouldn't make it. Pint after pint of blood flowed through their bodies, soon exhausting the supply on hand.

We had expected the bleeding, because we had to thin their blood with an anticoagulant in order to use the heart-lung machine. When we restarted their hearts the blood was effectively anticoagulated, and we faced intense bleeding in the area of the wound.

Their traumatized brains began to swell dramatically—which actually helped to seal off some of the bleeding vessels—but we didn't want it to cut off the blood supply.

The most harrowing moment came when we learned that the supply of blood might run out. Rogers called the hospital blood bank.

"I'm sorry, but we don't have much blood on hand," said the voice on the other end of the line. "We've checked and there's no more anywhere in the city of Baltimore."

"I'll give mine if you need it," someone said as soon as Mark Rogers reported.

Immediately six or eight people in the operating room volunteered on the spot to donate their blood, a noble gesture but one that wasn't practical. Finally the Hopkins blood bank called the American Red Cross, and they came through with ten units—exactly what we needed.

By the time the operation was over, the twins had used 60 units of blood—several dozen times more than their normal blood volume. The extensive head wounds measured approximately 16 inches in circumference.

While this was going on, someone from the team was staying in touch with the parents, who had left their hotel and were now in the waiting room. We also had staff on hand making sure those of us on the team had food to eat during our infrequent breaks.

We had planned to fit the twins immediately with Dufresne's creation of a titanium mesh covering mixed with a paste of crushed bone from the babies' shared portion of the skull. Once in place, the babies' skull bones would grow into and around the mesh, and it wouldn't need removal.

First, however, we had to be able to get their scalps closed before their swelling brains came completely out of their skulls. We put the boys into a barbiturate coma to slow down the metabolic rate to the brain. Then Long and I moved back, and Dufresne and his plastic surgery team went into action, working furiously trying to get the scalp back together. Finally they got things pretty much together on one boy with a few gaps on the other.

Dufresne would have to wait for a later date to install the titanium plates.[3]

We also ran into the problem that we didn't have enough scalp to cover both infants' heads; we temporarily closed Benjamin's with surgical mesh. Dufresne would plan a second operation to create a cosmetically acceptable skull if the infants continued to recover.

If the infants continued to recover.

---

[1] Siamese twins occur once in every 70,000 to 100,000 births; twins joined at the head occur only once in 2 to 2.5 million births. Siamese twins received their name because of the birthplace (Siam) of Chang and Eng (1811-1874) whom P.T. Barnum exhibited across America and Europe.

Most cranio pagus Siamese twins die at birth or shortly afterward. So far as we know, not more than 50 attempts had previously been made to separate such twins. Of those, less than ten operations have resulted in two fully normal children. Aside from the skill of the operating surgeons, the success depends largely on how much and what kind of tissue the babies share. Occipital cranio pugus twins (such as the Binders) had never before been separated with both surviving.

Other Siamese twins joined at the hip or chest had been done successfully. Even so, when any two children are born with their bodies together, an attempt to separate them is an extremely delicate operation with chances of survival normally no greater than fifty-fifty. The twins share certain biosystems and, if damaged, would result in both their deaths.

[2] On March 6, 1982, Alex Haller and a 21-member Johns Hopkins medical team had performed a successful separation of twin girls born to Carol and Charles Selvaggio of Salisbury, Massachusetts, in a ten-hour operation. Emily and Francesca Selvaggio were joined from the chest to the upper abdomen, sharing an umbilical cord, skin, muscle, and rib cartilage. Haller's team had their major problem with intestinal obstructions.

[3] Benjamin and Patrick would have to make another 22 trips into the operating room for the complete closure of their scalps. While I did a few of the operations, Dufresne did most of them, including some fancy flaps to cover the back of Benjamin's head.

20 THE REST OF THEIR STORY

I*f they recover.* In every phase of the surgery, this was the underlying question. If. *Oh, God,* I prayed silently again and again, *let them live. Let them make it.*

Even if they survived the surgery, weeks would lapse before we could fully assess their condition. The waiting would be a constant strain because we would be constantly looking for the first signs of normalcy, all the while fearing that we might detect signs of brain damage.

To give their severely traumatized brains a chance to recover without any lasting ill effect, we used the drug phenobarbitol to put the babies into an artificial coma. Phenobarbitol drastically reduced their brains' metabolic activity. We hooked them up to life-support systems that controlled their blood flow and respiration. The brain swell was severe, but not worse than we had expected. We indirectly monitored the swelling by measuring changes in heart rate and blood pressure and by periodic CT scans which give a three-dimensional X-ray picture of the brain.

The surgery ended at 5:15 a.m. on Sunday morning. It had taken 22 hours. And the battle wasn't over yet.

When our team emerged from surgery to the sound of the applause of other hospital staff members, Rogers went directly to Theresa Binder and, with a smile on his face, asked, "Which child would you like to see first?"

She opened her mouth to respond, and tears filled her eyes.

Once we set in motion the plan to separate the Binder twins, the public relations office at Johns Hopkins informed the media of what we were doing. This was a historic operation. Although we hadn't known it, the waiting room and corridors were alive with reporters. Naturally, none of them got into the operating room. Heavy security in the hospital would have stopped them even if they had tried to get inside. Several of the local radio stations gave updates on the surgery every hour. Naturally, with this kind of coverage untold thousands of the general public suddenly became involved in this surgical phenomenon. Later, I learned that many of the people who followed the updates had stopped during the day and prayed for our success.

Once out of the operating room, exhaustion took over, and we wanted to collapse. In the minutes after surgery, I couldn't think of answering anybody's questions or talking about what we had done. Rogers delayed a press conference until later that afternoon, giving us a chance to rest and clean up a little. At 4:00 when I walked into the conference room the magnitude of this surgery hit me. The room was wall-to-wall reporters with cameras and microphones. It may seem strange, but when one is doing a job—no matter what the job is—it's hard to comprehend the importance of it.

That afternoon, only a few hours out of surgery, my thoughts centered on Patrick and Benjamin Binder. The media attention the historic surgery generated was one of the last things on my mind. In fact, I doubt that any of us were prepared for the response of reporters and the myriad questions they asked. We must have looked strange standing in front of the media people, with our wrinkled clothes and fatigue-filled faces. We were tired but elated. The first step had been a giant one, and we'd made it. But it was only the first step on a long road.

"The success in this operation is not just in separating the twins," Mark Rogers said at the beginning of the news conference. "Success is producing two normal children."

As Rogers answered questions, I kept thinking how grateful I felt to have been a part of this magnificent team. For five months we had been one unit, all specialists and all tackling the same problem together. The staff at the pediatric ICU and the consultants in the children's center reacted spectacularly. They rallied behind us and spent countless hours without charge, working to make this operation successful.

I listened as Rogers explained the steps of the surgery and added, "It shook me that we were able to perform as a team at this level of complexity. We are capable of doing even better things than we believe we are, if we challenge each other to do it."

Although some of the others responded to questions, as the chief spokesmen Mark Rogers and I answered most of them. When reporters asked me about the boys' chances for survival, I told them, "The twins have a 50-50 chance. We had thought the whole procedure out well. Logically it ought to work, but I also know that when you do what hasn't been done before unexpected things are bound to happen."

One reporter raised the question about their vision, "Will they be able to see? Both of them?"

"At this point, we simply don't know."

"Why not?"

"Number one," I said, "the twins are too young to tell us themselves!" I did get a laugh from some of them. "Number two," I continued, "their neurological condition was impaired, and that would delay our ability to assess their visual capabilities. The boys were not yet capable of looking at things or following objects with their eyes."

(The next day all over the world, headlines blared, TWINS BLIND FROM SURGERY. We never said that or implied any such statement. We said we couldn't tell.)

"But will they survive?" asked a reporter.

"Can they live normal lives?" asked another.

"It's all in God's hands now," I said. Besides believing that statement, I didn't know what else to say. As I walked out of the crowded room, I realized I had said everything that needed to be said.

As pessimistic as I was about the eventual outcome of the surgery, I still felt a glow of pride in being able to work side by side with the best men and women in the medical field. And the end of the surgery wasn't the end of our teamwork. The postoperative care was as spectacular as the surgery. Everything in the weeks following the surgery confirmed again our togetherness. It seemed as if everyone from ward clerks to orderlies to nurses had become personally involved in this historic event. We were a team—a wonderful, marvelous team.

Patrick and Benjamin Binder remained in a coma for ten days. This meant that for a week and a half nobody knew anything. Would they remain comatose? Would they wake up to start living a normal life? Be handicapped? We all waited. And we wondered. Probably most of us worried a little and prayed a lot.

We hadn't done anything unusual by putting them into comas. We had put individuals in barbiturate comas for periods that long before. For instance, children with severe head trauma need the comas to keep their intracranial pressures down. We constantly checked the twins' vital signs, we felt the skin flaps to see how tense they were. Initially they were quite tense, and then they started to soften—a good sign that said the swelling was lessening. Occasionally when the barbiturate concentration would decrease, and we would see a movement, we'd say, "Well, they can move." At this point we needed every sign of hope.

"It's all in God's hands," I'd say, and then remind

myself, "That's where it's always been."

For at least the next week, whenever I'd go off duty I expected someone to call me and say, "Dr. Carson! One of the twins has had a cardiac arrest. We're resuscitating him now." I couldn't relax much at home either, because I just knew the phone would ring and I'd hear the terrible, dreaded message. It wasn't that I didn't trust God or our medical team. It's just that we were in uncharted waters and, as doctors, knew that the complications were endless. I always expected the bad news; fortunately, it never came.

In the middle of the second week, we decided to lighten up on the coma.

"They're moving," I said a couple of hours later when I stopped by to check. "Look! He moved his left foot! See!"

"They're moving!" someone beside me said. "They're both going to make it!"

We were beside ourselves with joy, almost like new parents who must explore every inch of their new babies. Every movement from a yawn to the wiggling of toes became a cause of celebration throughout the hospital.

And then came the moment that brought tears to many of us.

That same day, as soon as the phenobarbitol wore off, both boys opened their eyes and started looking around. "He can see! They can both see!" "He's looking at me! See—see what happens when I move my hand." We would have sounded crazy to anyone who didn't know the five-month history of preparation, work, worry, and concern. But we felt exhilarated. In the days that followed I'd find myself silently asking, *Is this real? Is this happening?* I hadn't expected them to survive for 24 hours, and they were progressing nicely every day. "God, thank You, thank You," I heard myself say again and again. "I know You have had Your hand in this."

We did have some post-operative emergencies but nothing that didn't come under control quickly. The pediatric anesthesiologists run the ICU. The people who had invested a tremendous amount of their time in this operation were the same ones who had been taking care of them postoperatively, so they really stayed on top of the situation.

Then questions arose about their neurological ability. What would they be able to do? Could they learn to crawl? Walk? To perform normal activities?

Week by week Patrick and Benjamin started doing more and more things and interacting more responsively. Patrick in particular got to the point where he was playing with toys, rolling from one side to the other and doing well with his feet. One day, however, about three weeks before he went back to Germany, Patrick unfortunately aspirated (sucked) his food into his lungs. A nurse discovered him in bed in respiratory arrest. Her quick thinking enabled an emergency team to resuscitate him, but no one knew how long he had been without breathing. He was already blue. He wasn't the same after that. Sadly, without saying it, we knew this meant some kind of brain damage, but we had no idea how extensive. The brain can't tolerate more than a few seconds without oxygen. At the time the twins left Johns Hopkins, Patrick, despite his respiratory arrest, was making strides. Benjamin continued to do quite well, even though his responses were slower at first. He was soon doing the things Patrick was doing before he had his arrest, such as rolling from side to back.

Unfortunately, because of the parents' contractural agreement with *Bunte* magazine, I can write nothing about the progress of the twins after they left Johns Hopkins. On February 2, 1989, I do know that two separated and much-loved twin boys celebrated their second birthdays.

21 FAMILY AFFAIRS

Candy's voice, near, urgent, called me from a deep sleep at 2:00 o'clock in the morning. "Ben! Ben! Wake up."

I burrowed deeper into my pillow. It had been a tiring day. I'd spent the day—May 26, 1985—at our church, involved in an event for runners called Healthy Choices. We had invited people to run one kilometer, five, or ten. Other doctors and I gave quick physical examinations and personal health profiles while experts provided tips on healthier living and better running.

Candy, waiting out her final month of pregnancy, had walked in the One K. Now she nudged me and said, "I'm having contractions."

I forced my eyes to crack open. "How far apart?"

"Two minutes."

It only took a moment for that message to leap into my brain. "Get dressed," I commanded as I leaped from the bed. We had a half-hour drive ahead of us to get her to Hopkins. Our first son, born in Australia, had come after eight hours of labor. We figured this one would arrive a little faster.

"The pains started just a few minutes ago," she said, swinging her feet to the floor and pulling herself out of bed. Halfway across the room, Candy paused. "Ben, they're coming more frequently." Her voice was so matter-of-fact she could have been commenting on the weather.

I don't recall what I answered. I was fairly calm, still methodically getting dressed.

"I think the baby's coming," Candy said. "Now."

"You're sure?" I jumped up, grabbing her shoulders and helping her back into bed. I could see that the head was starting to crown. She lay quietly and pushed. I felt perfectly fine and not particularly excited. Candy behaved as if she delivered a baby every other month. I recall being thankful for my experience in delivering babies, aware that they had all been brought into this world under better circumstances.

Within minutes I had caught the baby. "A boy," I said. "Another boy."

Candy tried to smile, and the contractions continued. I waited for the placenta. My mother was staying with us, and I yelled to her, "Mother, bring towels! Call 911!" Afterward I wondered if my voice sounded like it did with a four plus emergency.

Once I had the placenta, I said, "I need something to clip the umbilical cord. Where can I find something?" My main concern then was to clamp the umbilical cord, and I had no idea what to use.

Without answering me, Candy pulled herself out of bed and walked fairly steadily into the bathroom, returning immediately with a large bobby pin. I put it on the cord. About that time I heard the paramedics arriving. They took Candy and our newborn, whom we named Benjamin Carson, Jr., to the local hospital.

Later my friends asked, "Did you charge a delivery fee?"

———

"Too busy," I told myself for the hundreth time. "Something's got to change." It was an echo, a bouncing off the wall echo, that I'd repeated time and time before.

This time I knew I had to make changes.

Like others at Hopkins, I faced a serious dilemma with an active neurosurgical career. Working in a teaching hospital demanded a greater commitment to time and patients than I would have faced if I'd had my own practice. "How do I find adequate time to spend with my family?" I asked myself.

Unfortunately, neurosurgery is one of those unpredictable fields. We never know when problems are going to arise, and many of them are extremely complex, requiring a tremendous investment of time. Even if I devoted myself exclusively to a clinical practice, I would still have bad hours. When I throw on top of that the necessity of continuing laboratory research, writing papers, preparing lectures, remaining involved in academic projects, and more recently, presenting motivational talks to young people, there weren't enough hours in any day or week. It meant that if I wasn't careful, every area of my life would suffer.

For days I thought about my schedule, my commitments, my values, and what I could eliminate. I liked everything I was doing, but I saw the impossibility of trying to do it all. First, I concluded that my top priority was my family. The most important thing I could do was to be a good husband and father. I would reserve my weekends for my family.

Second, I wouldn't allow my clinical activities to suffer. I decided to go all out to be the best clinical neurosurgeon I could be and contribute as much as I could to the well being of my patients. Third, I wanted to serve as a good role model to young people.

Although I believe it was the correct decision, the process wasn't easy. It meant budgeting my time, giving up things I enjoyed doing, even things that would further my career. For instance, I'd like to do more publishing in the medical field, sharing what I've learned and pushing toward more intense research. Public speaking appeals to me, and more opportunities were coming my way to speak at national meetings. Naturally, these outlets also would enable me to advance rapidly through academic ranks. Fortunately many of those things seem to be happening anyway, but not as fast as they would if I were able to devote more time to them.

Important also was the need to spend time in my own church. Right now I'm an elder at Spencerville Seventh-day Adventist Church. I'm also Health and Temperance Director, which means I present special programs and coordinate the other medical workers in our church. For instance, we sponsor activities such as marathons, and I help in coordinating such events and organizing the medical screening. Our denomination stresses health, and I promote the health-conscious magazines *Vibrant Life* and *Health* among our congregation.

I also teach an adult Sabbath school class in which we discuss the issues of Christianity and their relevancy to our daily lives.

The first step toward freeing my time took place in 1985. We had gotten so busy at the hospital that we had to bring in another pediatric neurosurgeon. This additional staff member took some pressure off me. Hiring another man was quite a step for Hopkins because, since the beginning of the institution in the last century, pediatric neurosurgery had been a one-person department. Even today few institutions have two professionals on staff. At Hopkins we're talking about three, and possibly a fellowship in pediatric neurosurgery, because we have such a high volume of cases, and we see no signs of its abating.

Additional personnel didn't really solve my dilemma, however. Early in 1988 I admitted to myself that no matter how hard I worked or how efficiently, I would never finish the work, not even if I stayed in the hospital until midnight. Then I made my decision—one that, with God's help, I could stick to. I would leave for home every evening at 7:00 o'clock, 8:00 at the latest. That way I could at least see my children before they went to bed.

"I can't finish everything," I said to Candy, who has been totally supportive. "It's impossible. There's always just a little more to be done. So I may as well leave work unfinished at 7:00 p.m. instead of 11:00."

I've held to that schedule. I finish my work at the hospital by 7:30, and I'm back at the office 12 hours later. It's still a long day, but working 11 or 12 hours is reasonable for a doctor. Staying at it 14 to 17 hours isn't.

As more speaking opportunties come, they involve traveling. When I have to go a great distance, I take the family with me. When the children get into school that will have to change. For now, whenever I'm invited to speak, I ask if transportation and accommodations can be provided for my family too.

We're anticipating that my mother will be living with us soon, and she can take care of the children sometimes while Candy and I travel. As busy as I am, as many people as require my time, I think it will be good for Candy and me to be alone together. Without her support my life would not be the success it is today.

———

Before we married I told Candy that she wouldn't see much of me. "I love you, but I'm going to be a doctor, and that means I'm going to be very busy. If I'm going to be a doctor I'll be a driven person, and it's going to take a lot of time. If that's something you can live with then we can get married, but if you can't, we're making a mistake."

"I can deal with that," she said.

Did I sound selfish? Did my idealism cloud my commitment to the woman who would be my wife? Perhaps the answer is Yes on both questions, but I was also being realistic.

Candy has coped extremely well with my long hours. Maybe it's because she is confident and secure in herself that she can support me so well. Because of her support, I handle the demands more easily.

While I was an intern and a junior resident, I was seldom around because I worked 100 to 120 hours a week. Obviously, Candy seldom saw me. I'd call her, and if she had a few minutes she'd come over and bring my meal. I'd eat, and we'd spend a few minutes together before she went home.

During that period, Candy decided to return to school. She said, "Ben, I'm at home every night by myself so I may as well go and do something." Candy has a lot of creative energy, and she put it to use. At one church she started a choir, and an instrumental ensemble in another. During our year in Australia, she started a choir and instrumental ensemble.

We now have three children. Rhoeyce was born December 21, 1986, making us a family of five. I grew up without a father and I don't want my sons to grow up without one. It's vitally important that they know *me*, rather than just looking at my pictures in a scrapbook or magazine or seeing me on television. My wife, my sons—they are the most important part of my life.

22 THINK BIG

Candy and I share a dream, a dream unfulfilled as yet. Our dream is to see a national scholarship fund set up for young people who have academic talent but no money. This scholarship would help them to gain any type of education they want in any institution they want to attend. Most philanthropic funds are too politically oriented and depend too much on knowing the right people or getting important people behind you.

We dream of a scholarship program that recognizes *pure talent* in any field. We dream of seeking out those gifted young people who deserve a chance for success but would never be able to get near it because of lack of funds.

I would very much like to be in a position where I could do something to help make that dream a reality.

I put THINK BIG into practice in my own life. As my life moves forward, I want to see thousands of deserving people of every race moving into leadership because of their talents and commitments. People with dreams and commitments can make it possible.

"What's the key to your success?" the teenage boy with the Afro asked.

It wasn't a new question. I'd heard it so many times that I finally worked out an acrostic answer.

"Think big," I told him.

I'd like to break this down and explain the meaning of each letter.

THINK BIG

T = TALENT

Learn to recognize and accept your God-given talents (and we all have them). Develop those talents and use them in the career you choose.

Remembering *T* for talent puts you far ahead of the game if you take advantage of what God gives you.

T also = TIME

Learn the importance of time. When you are always on time, people can depend on you. You prove your trustworthiness.

Learn not to waste time, because time is money and time is effort. Time usage is also a talent. God gives some people the ability to manage time. The rest of us have to learn how. And we can!

H = HOPE

Don't go around with a long face, expecting something bad to happen. Anticipate good things; watch for them.

H also = HONESTY

When you do anything dishonest, you must do something else dishonest to cover up, and your life becomes hopelessly complex. The same with telling lies. If you're honest, you don't have to remember what you said the last time. Speaking the truth each time makes life amazingly simple.

I = INSIGHT

Listen and learn from people who have already been where you want to go. Benefit from their mistakes instead of repeating them. Read good books like the Bible because they open up new worlds of understanding.

N = NICE

Be nice to people—all people. If you're nice to people, they'll be nice to you. It takes much less energy to be nice than it does to be mean. Being kind, friendly, and helpful takes less energy and relieves much of the pressure.

K = KNOWLEDGE

Knowledge is the key to independent living, the key to all your dreams, hopes, and aspirations. If you are knowledgeable, particularly more knowledgeable than anybody else in a field, you become invaluable and write your own ticket.

B = BOOKS

I emphasize that active learning from reading is better than passive learning such as listening to lectures or watching television. When you read, your mind must work by taking in letters and connecting them to form words. Words make themselves into thoughts and concepts.

Developing good reading habits is something like being a champion weightlifter. The champion didn't go into the gym one day and start lifting 500 pounds. He toned his muscles, beginning with lighter weights, always building up, preparing for more. It's the same thing with intellectual feats. We develop our minds by reading, by thinking, by figuring out things for ourselves.

I = IN-DEPTH LEARNING

Superficial learners cram for exams but know nothing two weeks later. In-depth learners find that the acquired knowledge becomes a part of them. They understand more about themselves and their world. They keep building on prior understanding by piling on new information.

G = GOD

Never get too big for God. Never drop God out of your life.

I usually conclude my talks by telling young people,

"If you can remember these things, if you can learn to THINK BIG, nothing on earth will keep you from being successful in whatever you choose to do."

My concern for young people, especially disadvantaged young people, first hit me the summer I worked as a recruiter for Yale. When I saw the SAT scores of those kids and how few of them made anywhere near 1200, it saddened me. It also bothered me because I knew from my own experiences growing up in Detroit that scores didn't always reflect how smart people are. I had met a lot of bright youngsters who could grasp things quickly, and yet, for a variety of reasons, they scored poorly on their SAT exams.

"Something's wrong with a society," I've told Candy more than once, "that has a system precluding these people from achieving. With the right help and the right incentive, many disadvantaged kids could achieve outstanding results."

I made a commitment to myself that at every opportunity, I'd encourage young people. As I became more well-known and started getting more opportunities to speak, I decided that teaching kids how to set goals and achieve them would be a constant theme of mine. Nowadays I get so many requests, I can't accept anywhere near all of them. Yet I try to do as much as I can for young people without neglecting my family and my duties at Johns Hopkins.

I have strong feelings on the subject of American youth and here's one of them. I'm really bothered at the emphasis given by the media on sports in the schools. Far too many youngsters spend all their energies and time on the basketball courts, wanting to be a Michael Jordan. Or they throw their energies toward being a Reggie Jackson on the baseball diamond or an O.J. Simpson on the football field. They want to make a million dollars a year, not realizing how few who try make those kinds of salaries. These kids end up throwing their lives away.

When the media doesn't emphasize sports, it's music. I often hear of groups—and many of them good

—who pour out their hearts in a highly competitive career, not realizing that only one group in 10,000 is going to make it big. Rather than putting all their time and energy into sports or music, these kids—these bright, talented young people—should be spending their time with books and self-improvement, ensuring that they'll have a career when they're adults.

I fault the media for perpetuating these grandiose dreams. I spend quite a bit of time talking to the freshmen groups and trying to help them realize that they have a responsibility to each one of the communities they have come from to become the best they can be.

While going to schools and talking to these young people, I try to show them what they can do and that they can make a good living. I urge them to emulate successful adults in the various professions.

To the successful professionals I say, "Take young people to your house. Show them the car you drive, let them see that you have a good life too. Help them to understand what goes into getting that good life. Explain that there are many ways to a fulfilled life besides sports and music."

A lot of young people are terribly naive. I've heard one after another say, "I'm going to be a doctor," or "a lawyer," or maybe, "president of the company." Yet they have no idea what kind of work goes into achieving such positions.

I also talk to parents, teachers, and anybody else associated with the community, asking them to focus on the needs of these teens. These kids must learn how to achieve change in their lives. They need help. Otherwise things will never get better. They'll just get worse.

Here's an example of how this works. In May 1988 the Detroit *News* ran a feature story on me in their Sunday supplement. After reading the article, a man wrote to me. He was a social worker and had a 13-year-old son who also wanted to be a social worker. However, things had not been going well. The father had been evicted, then lost his job. He and his son were looking for their next meal and his world had turned upside down. He was so depressed that he was ready to commit suicide. Then he picked up the Detroit *News* and read the article. He wrote:

"Your story just turned my life around and gave me hope. Your example inspired me to go on and put my best efforts into life again. I now have a new job, and things are starting to turn around. That article changed my life."

I've also gotten a number of letters from students in various schools who were not doing well, but, through their reading about me, seeing me on television, or hearing me speak, were challenged to redouble their efforts. They're making an attempt to learn things and that means they're going to be the best they can be.

TEST
Directions: Check one

Question: Who do you admire most?

☐ Michael Jackson
☐ Larry Bird
☒ Dr. Ben Carson
☐ Barbara Walters
☐ Ronald Reagan

P.S. Anyone with "half a brain" could answer that!

Jimmie Hankey

A single-parent mother wrote, telling me she had two children, one of whom wanted to be a fireman, the other a doctor. She said they had all read my story and had been inspired. Learning about my life and how my mother helped me turn my life around, actually inspired her to go back to school. By the time she wrote to me, she had been accepted into law school. Her children had turned their grades around and were doing very well. Letters like that make me feel very good.

At Old Court Middle School in the Baltimore suburbs they've started the Ben Carson Club. To be a member, students have to agree that they will watch no more than three television programs each week, and they will read at least two books. When I visited that school, they did a unique thing. Club members had previously received biographical information about my life and they held a contest. The winners were those students who correctly answered the most questions about me. On my visit, the six winners came to the stage and answered questions about me and my life. I listened, amazed at how much they knew about me and humbled that my life had touched theirs.

It still seems unreal to me when I go places and people are excited to see me. While I don't fully understand, I realize that particularly for Black people in this country I represent something that many of them have never seen in their lifetimes—someone in a technical and scientific area who has risen to the top. I'm recognized for my academic and medical achievements instead of for being a sports star or an entertainer.

While this doesn't happen often, *it does happen*, reminding me that I'm not the one big exception. For instance, I have a friend named Fred Wilson who is an engineer in the Detroit area. He's Black, and the Ford Motor Company selected him as one of their top eight engineers worldwide.

He's incredibly bright and has done outstanding work, yet few know about his achievements. When I make public appearances, I like to think I'm holding up my own life and all of the others who've shown that being a member of a minority race doesn't mean being a minority achiever.

I tell a lot of the students that I talk to about Fred Wilson and other Black high achievers who just don't get media attention or have a high profile. When you're in a field like mine at a place like Johns Hopkins and you're putting out your best, it's hard to hide. Whenever any of us here do anything outstanding, the media finds out and the word spreads. I know a lot of people in other, less-glamorous fields, who have done significant things, but hardly anyone knows about them.

One of my goals is to make sure that teenagers learn about these highly talented individuals so they can have a variety of role models. When young people have good role models, they can change and set their sights toward higher achievements.

Another goal is to encourage teenagers to look at themselves and their God-given talents. We all have these abilities. Success in life revolves around recognizing and using our "raw material."

I'm a good neurosurgeon. That's not a boast but a way of acknowledging the innate ability God has given to me. Beginning with determination and using my gifted hands, I went on for training and sharpening of my skills.

To THINK BIG and to use our talents doesn't mean we won't have difficulties along the way. We will—we all do. How we view those problems determines how we end up. If we choose to see the obstacles in our path as barriers, we stop trying. "We can't win," we moan. *"They* won't let us win."

However, if we choose to see the obstacles as hurdles, we can leap over them. Successful people don't have fewer problems. They have determined that nothing will stop them from going forward.

Whatever direction we choose, if we can realize that every hurdle we jump strengthens and prepares us for the next one, we're already on the way to success.